Mathematical Diamonds

© 2003 by
The Mathematical Association of America (Incorporated)
Library of Congress Catalog Card Number 2002114177

ISBN 0-88385-332-9

Printed in the United States of America

Current printing (last digit):
10 9 8 7 6 5 4 3 2 1

The Dolciani Mathematical Expositions

NUMBER TWENTY-SIX

Mathematical Diamonds

Ross Honsberger
University of Waterloo

Published and distributed by
The Mathematical Association of America

THE
DOLCIANI MATHEMATICAL EXPOSITIONS

Published by
THE MATHEMATICAL ASSOCIATION OF AMERICA

Committee on Publications
GERALD L. ALEXANDERSON, Chair

Dolciani Mathematical Expositions Editorial Board
DANIEL J. VELLEMAN, Editor
DONNA L. BEERS
ROBERT BURCKEL
SUSAN C. GELLER
LESTER H. LANGE
WILLIAM S. ZWICKER

The DOLCIANI MATHEMATICAL EXPOSITIONS series of the Mathematical Association of America was established through a generous gift to the Association from Mary P. Dolciani, Professor of Mathematics at Hunter College of the City University of New York. In making the gift, Professor Dolciani, herself an exceptionally talented and successful expositor of mathematics, had the purpose of furthering the ideal of excellence in mathematical exposition.

The Association, for its part, was delighted to accept the gracious gesture initiating the revolving fund for this series from one who has served the Association with distinction, both as a member of the Committee on Publications and as a member of the Board of Governors. It was with genuine pleasure that the Board chose to name the series in her honor.

The books in the series are selected for their lucid expository style and stimulating mathematical content. Typically, they contain an ample supply of exercises, many with accompanying solutions. They are intended to be sufficiently elementary for the undergraduate and even the mathematically inclined high-school student to understand and enjoy, but also to be interesting and sometimes challenging to the more advanced mathematician.

---

1. *Mathematical Gems*, Ross Honsberger
2. *Mathematical Gems II*, Ross Honsberger
3. *Mathematical Morsels*, Ross Honsberger
4. *Mathematical Plums*, Ross Honsberger (ed.)
5. *Great Moments in Mathematics (Before 1650)*, Howard Eves
6. *Maxima and Minima without Calculus*, Ivan Niven
7. *Great Moments in Mathematics (After 1650)*, Howard Eves
8. *Map Coloring, Polyhedra, and the Four-Color Problem*, David Barnette
9. *Mathematical Gems III*, Ross Honsberger
10. *More Mathematical Morsels*, Ross Honsberger
11. *Old and New Unsolved Problems in Plane Geometry and Number Theory*, Victor Klee and Stan Wagon
12. *Problems for Mathematicians, Young and Old*, Paul R. Halmos
13. *Excursions in Calculus: An Interplay of the Continuous and the Discrete*, Robert M. Young
14. *The Wohascum County Problem Book*, George T. Gilbert, Mark Krusemeyer, and Loren C. Larson
15. *Lion Hunting and Other Mathematical Pursuits: A Collection of Mathematics, Verse, and Stories by Ralph P. Boas, Jr.*, edited by Gerald L. Alexanderson and Dale H. Mugler
16. *Linear Algebra Problem Book*, Paul R. Halmos
17. *From Erdős to Kiev: Problems of Olympiad Caliber*, Ross Honsberger
18. *Which Way Did the Bicycle Go? ...and Other Intriguing Mathematical Mysteries*, Joseph D. E. Konhauser, Dan Velleman, and Stan Wagon
19. *In Pólya's Footsteps: Miscellaneous Problems and Essays*, Ross Honsberger
20. *Diophantus and Diophantine Equations*, I. G. Bashmakova (Updated by Joseph Silverman and translated by Abe Shenitzer)

21. *Logic as Algebra,* Paul Halmos and Steven Givant
22. *Euler: The Master of Us All,* William Dunham
23. *The Beginnings and Evolution of Algebra,* I. G. Bashmakova and G. S. Smirnova (Trans. by Abe Shenitzer)
24. *Mathematical Chestnuts from Around the World,* Ross Honsberger
25. *Counting on Frameworks: Mathematics to Aid the Design of Rigid Structures,* Jack E. Graver
26. *Mathematical Diamonds,* Ross Honsberger

MAA Service Center
P. O. Box 91112
Washington, DC 20090-1112
1-800-331-1MAA   fax: 301-206-9789

Preface

This miscellaneous collection of elementary gems contains brilliant insights from many fine mathematical minds. The majority of its more than 75 topics come from Euclidean geometry, combinatorial geometry, algebra and number theory, and most of the discussions can be followed comfortably by a college freshman.

There is no attempt to give instruction; in the few places where preliminaries are presented it is done only in preparation for a gem to follow.

The essays, written in a leisurely style, are intended as mathematical entertainment. While a measure of concentration is the price of enjoying some of these explanations, it is hoped that they will combine the excitement and richness of elementary mathematics with reading pleasure. Although it is not necessary to try the problems before going on to the solutions, if you are able to give them a little thought first, I'm sure you will find them all the more exciting.

The sections are independent and may be read in any order.

A few exercises (without solutions) are sprinkled throughout the book and a set of fifteen miscellaneous challenges (with solutions) is given at the end for your further enjoyment.

The book concludes with an index of publications, a subject index and a general index.

I would like to take this opportunity to thank Professor Dan Velleman and the members of the Dolciani Editorial Board for their warm reception and gentle criticism of the manuscript. The book is much improved because of their dedication and I am deeply grateful to them. It is again a pleasure to extend my warmest thanks to Elaine Pedreira and Beverly Ruedi for their unfailing geniality and technical expertise in seeing the manuscript through publication.

*God must love mathematicians—
He's given us so much to enjoy!*

Contents

Preface		vii
Section 1.	The Remarkable Wine-Rack Property	1
Section 2.	The Restless Lion	5
Section 3.	Apples and Sticks	9
Section 4.	A Surpising Result of Paul Erdős	13
Section 5.	Miscellaneous Gleanings	19
Section 6.	An Application of Turán's Theorem	35
Section 7.	Four Problems from Putnam Papers	43
Section 8.	Topics Based on Problems from *Quantum*	49
Section 9.	Two Distinguished Integers	79
Section 10.	A Property of the Binomial Coefficients	83
Section 11.	Nine Miscellaneous Problems	87
Section 12.	A Problem in Coin-Tossing	103
Section 13.	Semi-regular Lattice Polygons	109
Section 14.	Six Problems from the Canadian Open Mathematics Challenge	117
Section 15.	Three Pretty Theorems in Geometry	125
Section 16.	Two Gems from Euclidean Geometry	135
Section 17.	The Thue–Morse–Hedlund Sequence	143
Section 18.	Two Miscellaneous Problems	151
Section 19.	A Surprising Property of Regular Polygons	157
Section 20.	Three Short Stories in Number Theory	163
Section 21.	Three Geometry Problems	169
Section 22.	Three Problems From The 1990 Balkan Olympiad	175
Section 23.	A Japanese "Fan" Problem	181
Section 24.	Slicing a Doughnut	187
Section 25.	A Problem from the 1980 Tournament of the Towns	195

More Challenges 203
Solutions To The Challenges 209
Indices 237

SECTION 1
The Remarkable Wine-Rack Property

This fascinating discovery is the subject of problem 44 in the exciting collection of elementary problems *"Which Way Did The Bicycle Go?"* by Joe Konhauser, Dan Velleman, and Stan Wagon (MAA Dolciani Series, Vol. 18, 1996).

Across the horizontal bottom of rectangular wine rack $PQRS$ there is room for more than three bottles (A, B, C) but not enough for a fourth bottle (Figure 1). All the bottles that are put into this rack are the same size. Naturally, bottles A and C are laid against the sides of the rack and a second layer, consisting of just two bottles (D, E), holds B in place somewhere between A and C. Now we can lay in a third row of three bottles (F, G, H), with F and H resting against the sides of the rack. Then a fourth layer is held to just two bottles (I, J).

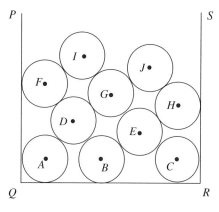

FIGURE 1

Now, if the bottles are not evenly spaced in the bottom row, the second, third, and fourth rows can slope considerably, tilting at different angles for different spacings. Prove, however, that whatever the spacing in the bottom row,

1

the fifth row is always perfectly horizontal!

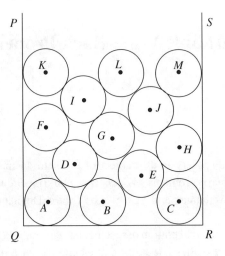

FIGURE 2

Proof: In Figure 3, KF is vertical, and so we need to show that $\angle FKL = 90°$.

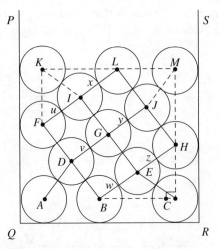

FIGURE 3

Clearly, the distance between the centers of touching bottles is just the diameter of a bottle. Thus I is equidistant from F, K and L, making I the circumcenter of $\triangle FKL$. Now, if I were to lie on FL, then FL would be a diameter of the circumcircle of $\triangle FKL$, making $\angle FKL$ the desired right angle. Thus it remains only to show that I lies on FL.

At the opposite part of the figure, we know that $\triangle BCH$ is right-angled at C and that E is its circumcenter, and so E is the midpoint of BH.

Now, clearly the four quadrilaterals around G are rhombuses, and hence parallelograms, making opposite sides u, v, and w parallel. Similarly, opposite sides x, y, and z are parallel. Therefore u and x, i. e., FI and IL, both lie in the direction of BEH, making them parts of the same straight line, and we are done already.

Similarly $\angle LMH$ is a right angle and the desired conclusion follows.

This property was discovered by Charles Payan of the Laboratoire de Structures Dèscretes et Didactique in France, one of the creators of the remarkable drawing package CABRI, while experimenting with this program. The solution is due to Hung Dinh.

SECTION 2
The Restless Lion

This gem comes from the wonderful collection *Mathematical Miniatures* by Svetoslov Savchev and Titu Andreescu (The Anneli Lax New Mathematical Library Series, MAA).

One day a restless lion roamed about in his circular cage, of radius 10 meters, along a polygonal path of total length 30 kilometers (Figure 1). Thus, coming to the end of one segment in his path, he would turn and proceed along the next segment in the path. Prove that, counting both turns to the left and turns to the right (considering each angle to be a positive quantity), he must have turned through a total of at least 2998 radians. (How can they say there's no humor in mathematics!)

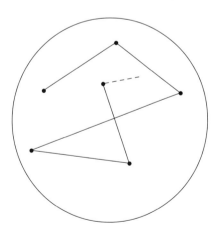

FIGURE 1

Let the path be $A_0 A_1 A_2 \ldots A_k$ and let the angle through which the lion turned at vertex A_i have magnitude φ_i radians (Figure 2). Now, it is conceivable that he could spin around on the spot or turn through an angle of $(2\pi - \varphi)$

in the other direction in order to effect a smaller turn than φ. However, let φ_i be the smallest angle by which the turn can be made. Thus no φ_i exceeds a straight angle, and the smallest total angle through which he could have turned is $T = \varphi_1 + \varphi_2 + \cdots + \varphi_{k-1}$. Any superfluous turning only increases this total.

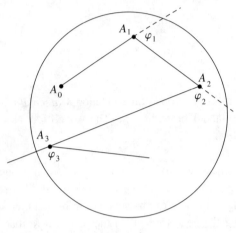

FIGURE 2

Extending segment $A_0 A_1$ into the ray R (Figure 3), let the path be straightened out into a 30-km segment $A_0 A'_k$ by unfolding it along R as follows:

Since R begins with $A_0 A_1$, let the rest of the path, $A_1 A_2 \ldots A_k$, be rotated rigidly about A_1 back through the angle φ_1 to bring $A_1 A_2$ into position $A_1 A'_2$ along R. Now, if this rotation takes A_3 to an intermediate position P_3, then the angle φ_2 would be carried to lie between R and $A'_2 P_3$. Therefore a further rigid rotation about A'_2 of the path beyond A'_2, through the angle φ_2 in the appropriate direction, would carry A_3, in the guise of P_3,

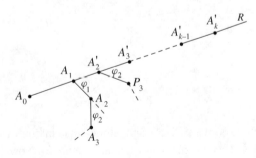

FIGURE 3

to a final image A'_3 on R. Clearly, continuing to unfold the path in this way about centers $A'_3, A'_4, \ldots, A'_{k-1}$, eventually yields a segment $A_0 A'_k$ of length 30 km.

One might wonder what the point of all this is. The brilliance of the approach lies in now considering how the center O of the cage is moved about by these rotations. Suppose the rotation $A_1(\varphi_1)$ carries O to O_1 (Figure 4). Then $\triangle O A_1 O_1$ is isosceles with vertical angle φ_1 and the altitude $A_1 M$ not only bisects the base OO_1 but also angle φ_1. Hence

$$OM = OA_1 \sin \frac{\varphi_1}{2}$$

and

$$OO_1 = 2OA_1 \sin \frac{\varphi_1}{2}.$$

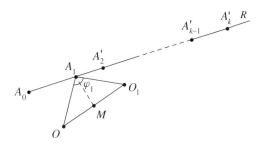

FIGURE 4

Now, since $\sin x < x$ for every $x > 0$, we have $\sin \varphi_i/2 < \varphi_i/2$. Moreover, OA_1 cannot exceed the radius 10 meters of the cage, and so

$$OO_1 = 2OA_1 \sin \frac{\varphi_1}{2} < 2 \cdot 10 \cdot \frac{\varphi_1}{2} = 10\varphi_1 \text{ meters.}$$

Similarly, the rotation about A'_2 carries O_1 to a point O_2 such that $O_1 O_2 < 10\varphi_2$ meters, and subsequent rotations carry the successive images of O through a sequence of points $O_3, O_4, \ldots, O_{k-1}$ that determine a second polygonal path $OO_1 O_2 \ldots O_{k-1}$ in which each segment $O_{i-1} O_i < 10\varphi_i$

meters (Figure 5). Hence the polygonal form of the triangle inequality gives

$$OO_{k-1} \leq OO_1 + O_1O_2 + \cdots + O_{k-2}O_{k-1}$$
$$< 10(\varphi_1 + \varphi_2 + \cdots + \varphi_{k-1}),$$

that is,

$$OO_{k-1} < 10T.$$

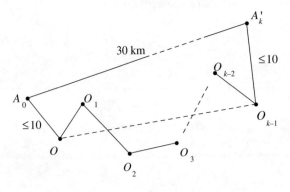

FIGURE 5

Now, since A'_k and O_{k-1} are both in the final image of the cage, and O_{k-1} is its center, the distance $O_{k-1}A'_k$ cannot exceed the radius 10 meters and we have in quadrilateral $A_0OO_{k-1}A'_k$ that

$$A_0A'_k \leq A_0O + OO_{k-1} + O_{k-1}A'_k$$
$$\leq 10 + OO_{k-1} + 10$$
$$= OO_{k-1} + 20.$$

Finally, since $A_0A'_k = 30000$ meters and $OO_{k-1} < 10T$, we obtain

$$30000 < 10T + 20$$

and the desired

$$2998 < T.$$

SECTION 3
Apples and Sticks*

1. Who Stole the Apples?

I expect many of us have spent time during our holidays working on some of those logical brainteasers that call for "each man's occupation" or "who is married to whom?" I suspect we find it hard to believe that the apparently irrelevant information that is given is actually enough to nail down the answer and our curiosity gets the better of us. At the early age that one is likely to be introduced to such a pastime, a straightforward process of elimination is probably the only approach at one's disposal. However, certain problems of this sort have very ingenious solutions. Consider the following lovely problem.

Out of six boys, exactly two were known to have been stealing apples. But Who?

Harry said "Charlie and George."
James said "Donald and Tom."
Donald said "Tom and Charlie."
George said "Harry and Charlie."
Charlie said "Donald and James."
Tom couldn't be found.

Four of the five boys interrogated had named one of the miscreants correctly and lied about the other one.
The fifth boy had lied outright!
Who stole the apples?

Let a graph A be constructed with a vertex for each boy—C for Charlie, G for George, and so on—and with an edge joining each pair of boys named

---

*This is a reworked version of material from my "Mathematical Gems" column in the *Two-Year College Mathematics Journal*, Vol. 10, January, 1979.

in the statements—the edge CG comes from Harry's statement "Charlie and George," and so forth.

Since four of the boys named one miscreant and one innocent person, four of the edges join a thief to an innocent person, and since one of the boys lied outright, the remaining edge has both ends at innocent parties. Hence the culprits altogether claim four of the endpoints of the edges and the innocents claim six. Since nobody told the whole truth, we observe that the two thieves are not joined by an edge.

Thus, in order to identify the thieves, we need to find two vertices in A which are not joined by an edge and which together claim a total of four of the endpoints. A brief look at A is sufficient to see that the only possibilities are C and J. Thus it was Charlie and James who stole the apples!

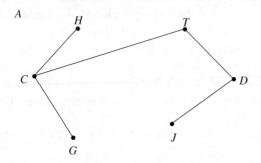

2. The Sticks

Suppose two pencil marks, of zero width, are made at random on a stick (that is, the marks are chosen independently and uniformly) and the stick is broken into k equal parts. What is the probability that the two marks are on the same piece of stick?

Of course, we don't have to break the stick; it is just as good to imagine that the stick has been marked off into k equal sections. Clearly, then, wherever the first pencil mark might happen to fall, the probability of having the second one in the same section is simply $1/k$.

With this little warmup, consider the following nice problem that was given to me by my colleague Steve Brown.

> Suppose two pencil marks are made at random on a stick which is then broken into k parts *at random*. What is the probability that the two marks are on the same piece of stick?

This time there is little chance of the pieces being all the same length. What do you think, then: is the probability greater than, less than, or the same as it was in the warmup problem? It might come as a mild surprise to learn that it is almost twice as likely for the marks to be on the same piece of stick in the present case. We can argue as follows.

The two pencil marks and the $k - 1$ break-points together form a set of $k + 1$ points on the stick. All these points are chosen at random and so the difference between a pencil mark and a break-point is purely conceptual. There are $\binom{k+1}{2}$ ways that two of a given set of $k + 1$ points on the stick can be specified to be the pencil marks, and so the same set of $k + 1$ points gives rise to $\binom{k+1}{2}$ cases. In each of these cases, the pencil marks occur on the same piece of stick if and only if they are *consecutive* along the stick, that is, in one of the k pairs of positions $(1, 2), (2, 3), (3, 4), \ldots, (k, k+1)$. Thus, for a given set of $k + 1$ points, the pencil marks will be on the same piece of stick in k of its $\binom{k+1}{2}$ cases, for a probability of

$$\frac{k}{\binom{k+1}{2}} = \frac{k}{\frac{(k+1)(k)}{2}} = \frac{2}{k+1}.$$

That is to say, no matter how the stick might be marked and broken, the probability of having the pencil marks on the same piece of stick is $\frac{2}{k+1}$.

SECTION 4
A Surpising Result of Paul Erdős*

In 1945, Paul Erdős directed attention to sets of points in the plane having the property that the distance between each pair of the points is an integer. Clearly any collection of integer points on the real line has this property, showing that such a set can contain any number of points, including an infinite number. Erdős's result is the surprising fact that essentially no other infinite set has this property:

> *if an infinite set of points in the plane determines only integer distances, then all the points lie on a straight line.*

The second surprise is that the proof centers around, of all things, *hyperbolas!*

(a) He proceeds indirectly. Suppose an infinite set S determines only integer distances and that some three of its points, A, B, and C, are not collinear. As usual, let a, b, and c, respectively, denote the lengths of the sides of this nondegenerate triangle which are opposite vertices A, B, and C. Let D be any other point of S. Not being a vertex, D can lie on the line of at most one side of $\triangle ABC$, implying that there are at least two sides on which it does not lie, say the lines AC and BC (Figure 1).

Let us begin by considering the consequences of not lying on BC. Because S determines only integer distances, the lengths of BD and CD are both integers and therefore the difference $|BD - CD|$ is an integer. Now, the triangle inequality applied to $\triangle DBC$ yields

$$|BD - CD| < BC = a,$$

which means that $|BD-CD| = k$, where k is one of the integers $0, 1, \ldots, a-1$. Accordingly, D is a point on the hyperbola which has foci at B and C and constant difference k between its focal radii. We can be no more specific than

---

*This is a reworked version of material from my "Mathematical Gems" column in the *Two-Year College Mathematics Journal*, Vol. 10, November, 1979.

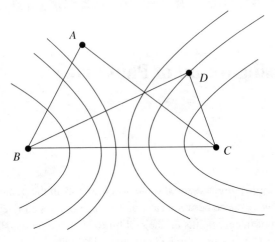

FIGURE 1

to say that D lies on one of the a hyperbolas of the family which have foci B and C and constant differences $\{0, 1, 2, \ldots, a-1\}$.

Similarly, not lying on AC, the point D must occur on one of the b hyperbolas which have foci at A and C and constant differences $\{0, 1, \ldots, b-1\}$. However, two different hyperbolas can intersect in at most four points, implying that D must be one of the (at most) $4ab$ points of intersection of the hyperbolas of these families. That is to say, there exist only a finite number of possible locations for any other point D of S, giving the contradiction that S cannot be infinite.

(b) We are thus forced to abandon the idea of an infinite set of noncollinear points which determines only integer distances. We might wonder, however, how far we can go with finite sets before having to take refuge in collinearity. We shall see that there are finite sets that determine only integer distances which are as large as you please; in fact, there exist such finite sets with an arbitrary number of points in which

no three points are collinear!

A *primitive* Pythagorean triple (a, b, c) is a triple of positive integers in which a, b, and c are relatively prime in pairs: that is, (a, b, c) satisfies $a^2 + b^2 = c^2$, where $(a, b) = (b, c) = (c, a) = 1$. There is a formula for such triples which provides any number of different triples. (Pythagorean triples are discussed briefly in chapter 13 of my *Ingenuity in Mathematics*, Vol. 23, NML Series, MAA, 1970.)

For any positive integer n, let (a_i, b_i, c_i), $i = 1, 2, \ldots, n-1$, denote $n-1$ different primitive Pythagorean triples. Denoting by P the product of the a_i, David Silverman showed in 1963 that only integer distances are determined by the following set of n non-collinear points in a coordinate plane:

$$(0, P) \quad \text{and} \quad \left(\frac{b_i}{a_i} P, 0\right), \quad i = 1, 2, \ldots, n-1.$$

Since a_i divides P, each of the $n-1$ points $(\frac{b_i}{a_i} P, 0)$ is an integer point on the x-axis, implying only integer distances among themselves. And the distance from $(0, P)$ on the y-axis to $(\frac{b_i}{a_i} P, 0)$ is

$$\sqrt{\frac{b_i^2 P^2}{a_i^2} + P^2} = P\sqrt{\frac{b_i^2}{a_i^2} + 1} = P\sqrt{\frac{b_i^2 + a_i^2}{a_i^2}} = P\sqrt{\frac{c_i^2}{a_i^2}} = \frac{Pc_i}{a_i},$$

which is an integer because a_i divides P. It remains only to show that the points $(\frac{b_i}{a_i} P, 0)$ are all different.

This follows nicely from the primitive character of the triples. Proceeding indirectly, suppose that the ratios $\frac{b_i}{a_i}$ and $\frac{b_j}{a_j}$ from two *different* triples (a_i, b_i, c_i), (a_j, b_j, c_j) are equal, making their corresponding points the same. Then $a_i b_j = a_j b_i$, and since a_i and b_i are relatively prime, a_i must divide a_j. Again, since a_j and b_j are relatively prime, a_j must divide a_i, and it follows that $a_i = a_j$. Similarly $b_i = b_j$, and we have the contradiction that the triples are not different.

(c) Silverman's set just escapes collinearity by the narrowest of margins. Let us conclude our story with the construction of an integer-set of n points which is as far removed from collinearity as possible—a set in which no three of its points are collinear. This is given in the outstanding book *Combinatorial Geometry in the Plane*, by Hadwiger, Debrunner, and Klee (Holt, Rinehart, and Winston, 1964, pages 5 and 6).

In the 3-4-5 right triangle, let θ denote the angle opposite the side of length 3. Then $\cos \theta = \frac{4}{5}$ and $\sin \theta = \frac{3}{5}$. Now, on the unit circle, centered at the origin, mark the points $P_0, P_1, P_2, \ldots$ around the circumference, starting at $P_0 = (1, 0)$, by repeatedly rotating the radius counterclockwise through an angular step of 2θ. Thus

$$P_n = e^{2n\theta i} = \cos 2n\theta + i \sin 2n\theta, \quad n = 0, 1, 2, \ldots,$$

implying that P_n is the point $(\cos 2n\theta, \sin 2n\theta)$ in the Cartesian plane. Now, there is a theorem that states

if $0 < \theta < \pi/2$ and $\cos\theta$ is rational, then
either $\theta = \pi/3$ or θ/π is irrational.

Its proof would carry us too far afield, but you can find a proof on page 58 of *Combinatorial Geometry in the Plane*. From this theorem it follows that all the points P_n are distinct:

Since $\cos\theta = 4/5$, $\cos\theta$ is rational and θ is not $\pi/3$. Proceeding indirectly, suppose P_n and P_m, $m \neq n$, were to coincide. In this case, their arguments would differ by an integral multiple of 2π, $2n\theta - 2m\theta = 2\pi q$ for some integer q, giving the contradiction

$$\frac{\theta}{\pi} = \frac{q}{n-m},$$

a rational number.

Thus, for each positive integer n, there exist n distinct points P_i on the circle, and the nonzero distance between two of them is

$$\begin{aligned}
P_s P_t &= \sqrt{(\cos 2s\theta - \cos 2t\theta)^2 + (\sin 2s\theta - \sin 2t\theta)^2} \\
&= \sqrt{2 - 2\cos 2s\theta \cos 2t\theta - 2\sin 2s\theta \sin 2t\theta} \\
&= \sqrt{2 - 2\cos[2(s-t)\theta]} \\
&= \sqrt{2 - 2[2\cos^2(s-t)\theta - 1]} \\
&= \sqrt{4 - 4\cos^2(s-t)\theta} \\
&= 2\sqrt{\sin^2(s-t)\theta} \\
&= 2|\sin(s-t)\theta|.
\end{aligned}$$

There is no guarantee that this distance is an integer, but it is easy to see that it is at least rational. From de Moivre's theorem and the binomial theorem we have

$$\begin{aligned}
&\cos k\theta + i \sin k\theta \\
&= (\cos\theta + i\sin\theta)^k \\
&= \cos^k\theta + \binom{k}{1}(\cos^{k-1}\theta)(i\sin\theta) + \binom{k}{2}(\cos^{k-2}\theta)(-\sin^2\theta) \\
&\quad + \binom{k}{3}(\cos^{k-3}\theta)(-i\sin^3\theta) + \binom{k}{4}(\cos^{k-4}\theta)(\sin^4\theta) + \cdots,
\end{aligned}$$

that is,

$$\cos k\theta + i \sin k\theta$$
$$= \left[\cos^k \theta - \binom{k}{2}\cos^{k-2}\theta \sin^2\theta + \binom{k}{4}\cos^{k-4}\theta \sin^4\theta - + \cdots\right]$$
$$+ i\left[\binom{k}{1}\cos^{k-1}\theta \sin\theta - \binom{k}{3}\cos^{k-3}\theta \sin^3\theta + - \cdots\right].$$

Equating imaginary parts, and recalling that $\cos\theta = \frac{4}{5}$ and $\sin\theta = \frac{3}{5}$, we have that $\sin k\theta$ is rational. Hence, for $k = s - t$, the distance

$$P_s P_t = 2\left|\sin(s-t)\theta\right|$$

is rational.

Thus the points $P_0, P_1, \ldots, P_{n-1}$ constitute a finite set of n distinct points which determine only rational distances. However, rational distances are just as good as integer distances! Let N denote a common denominator of the ratios which give these rational distances $P_i P_j$, and let the points P_i be projected from the origin to image points P_i' on a concentric circle C of radius N. The lengths $P_i' P_j'$ of the chords in C are therefore N times the corresponding chords $P_i P_j$ in the unit circle, making them all integers. And since the points P_i' lie on a circle, no three of them are collinear.

SECTION 5
Miscellaneous Gleanings

Many fine books of elementary gems and problems have been published recently. In this chapter we consider a few of the treasures from four of these books. Topics based on problems from *Quantum* appear in a separate essay.

1. The Mathematical Visitor

The *Mathematical Visitor* is an American journal that was published from 1877 to 1896. Stanley Rabinowitz has done the mathematics community an enormous service by publishing the 300 problems and solutions from this remarkable periodical. His book is appropriately entitled *Problems and Solutions from the Mathematical Visitor 1877–1896*, and may be ordered from The MathPro Press, P.O. Box 2039, Mansfield, OH 44905.

1. (Problem 273, by J. M. Quiroz, 1881)

It is well known that the formula $\triangle = rs$ gives the area of a triangle in terms of its inradius and semiperimeter. This problem concerns the adjustment that needs to be made in s if r is changed to the circumradius R.

Establish the formula $\triangle = Rs$, where s is the semiperimeter of the **orthic triangle**. (Recall that the vertices of the orthic triangle are the feet of the altitudes of the given triangle).

Two very nice straightforward proofs are given in *The Mathematical Visitor* and one doesn't wish for more. However, the result is a simple corollary to Leopold Fejér's brilliant solution of Fagnano's problem (see Coxeter's monumental *Introduction to Geometry*, Wiley, 1961, page 20, and *The Enjoyment of Mathematics* by Rademacher and Toeplitz, Princeton University Press, 1957, page 30, one of my all-time favorite books).

Let us begin with a review of Fejér's approach. Let the orthic triangle be *DEF* (Figure 1). The right angles at E and F make *FBCE* a cyclic quadrilateral

19

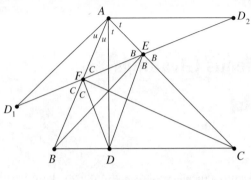

FIGURE 1

and hence the exterior angle *AEF* is equal to the interior angle *B* at the opposite vertex. Similarly, *ABDE* is cyclic and exterior angle *DEC* is also equal to angle *B* at the opposite vertex.

Now let *D* be reflected in *AC* to give D_2. Since this reflection duplicates $\angle DEC$ as $\angle CED_2$, angle *B* occurs three times at *E*, as in Figure 1. Thus angle FED_2 is equal to straight angle *AEC*, implying that the image D_2 is on *FE*. Similarly, reflecting *D* in *AB* takes *D* to D_1 on *EF* extended, making D_1FED_2 a straight segment whose length is obligingly the perimeter of the orthic triangle.

Now, each of these reflections duplicates part of angle *A* with the result that $\angle D_1 A D_2 = 2A$. Moreover, each of $D_1 A$ and $D_2 A$ is the image of the altitude *AD*. Hence (Figure 2)

$D_1 A D_2$ is isosceles with arms of length *AD*, its vertical angle is $2A$, and its base is equal to $2s$, the perimeter of the orthic triangle.

Thus the altitude *AX* bisects both the angle at *A* and the base $D_1 D_2$. Observing in $\triangle ABC$ that $AD = b \sin C$, we obtain

$$s = D_1 X = D_1 A \sin A = AD \sin A.$$

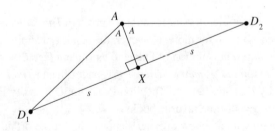

FIGURE 2

Since $R = \frac{a}{2\sin A}$, it follows that
$$Rs = \frac{a}{2\sin A} AD \sin A = \frac{1}{2} a \cdot AD,$$
which is one-half the base times the height of $\triangle ABC$, completing the proof.

2. (Problem 266, by W. F. L. Sanders, 1881)

Two boys, John and George, twice ran a footrace of 200 yards. In the first heat, John gave George a headstart of eight yards and two seconds running time and then proceeded to beat him by two seconds anyway. The next time, he gave George 16 yards and five seconds, but this was too much and George beat him by 20 yards. How fast do the boys run?

Let George's speed be u yards/sec and John's speed be v yards/sec. A nice way of getting two equations in u and v is to determine, for each heat, the length of time the boys were simultaneously engaged in running, that is, from the time John finally started running to the time the first one finished.

Thus, in the first heat, George is down the track at B by the time John starts at E (Figure 3), but John covers the entire 200 yards EF while George only makes it as far as C, where the remaining distance CD would take George another two seconds.

Now, since it takes George $\frac{192}{u}$ seconds to go the 192 yards from A to D, it only takes him $t_{BC} = \frac{192}{u} - 4$ seconds to go from B to C. Since this is all the time John needs to cover the whole course, our first equation is $t_{BC} = t_{EF}$:
$$\frac{192}{u} - 4 = \frac{200}{v}.$$

George •--8 yd--A--2 sec.--+-----B-----------C--2 sec--D•

|————————— 192 yd —————————|

$$t_{BC} = \frac{192}{u} - 4 \text{ sec}$$

John •E————————————————F•

200 yd

$$t_{EF} = \frac{200}{v} \text{ sec}$$

FIGURE 3

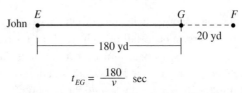

FIGURE 4

Similarly, in the second heat (Figure 4), George is at B when John starts at E, and George reaches D when John is only at G, which is 20 yards short of the finish line. Thus our second equation $t_{BD} = t_{EG}$ is

$$\frac{184}{u} - 5 = \frac{180}{v}.$$

Solving we easily find that George runs at a rate of $u = 8$ yds/sec and John at a very respectable $v = 10$ yds/sec.

3. (Problem 160, by L. P. Shidy, 1880)

Determine the location of a point O inside a given convex quadrilateral $ABCD$ such that the area of the quadrilateral is quartered by the segments from O to the midpoints of the sides (Figure 5).

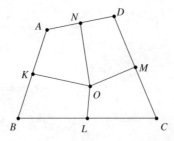

FIGURE 5

We follow the solution due independently to each of K. S. Putnam and John I. Clark.

Construct parallelogram $PQRS$ by drawing through A and C lines parallel to diagonal BD and through B and D lines parallel to diagonal AC (Figure 6).

Then the required point O is simply the center of $PQRS$, that is, the point of intersection of the diagonals PR and QS.

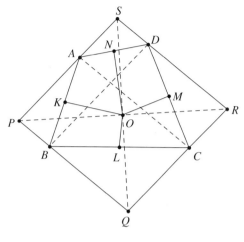

FIGURE 6

Any line through the center of a parallelogram bisects its area. Let UOV be a line through the centre O that is parallel to the side PS, thus making $PUVS$ itself a parallelogram with half the area of $PQRS$. For definiteness, suppose BD is on the side of UV as shown in Figure 7 (the argument is slightly different if D is on the other side of UV). It is now easy to see that the area of the typical region $KONA$ is one-quarter $ABCD$.

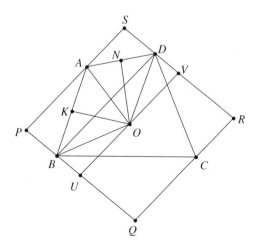

FIGURE 7

The proof is simply a matter of observing a few bisections. Recall that the area of a triangle is bisected by a median and that the area of a parallelogram is bisected by any triangle which has the same base and has its third vertex on the opposite side.

Thus $\triangle ABD$ bisects $PBDS$ and $\triangle BCD$ bisects $BQRD$. Hence $ABCD = PUVS$, each being one-half $PQRS$.

As just observed, $\triangle ABD$ bisects $PBDS$, and since $\triangle BOD$ bisects $BUVD$, it follows that $ABOD$ bisects $PUVS$. Thus

$$ABOD = \frac{1}{2}PUVS = \frac{1}{2}ABCD.$$

Finally, since medians OK and ON bisect triangles ABO and AOD, we have

$$KONA = \frac{1}{2}ABOD = \frac{1}{4}ABCD.$$

4. (Problem 272, by Benjamin Headley, 1881; solution due to E. B. Seitz, and to W. P. Casey)

How can a 2×10 rectangle R be cut into four pieces that can be reassembled into a square S?

One's first thought is likely to be that, since the area of R is $2 \cdot 10 = 20$, the side of S is $\sqrt{20} = 2\sqrt{5}$. Thus half the length of a side is $\sqrt{5}$, and the theorem of Pythagoras applied to $\triangle ABC$ in Figure 8 gives the distance AC from a vertex to the midpoint of an opposite side of S to be 5. Hence DE is also 5.

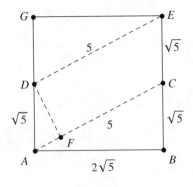

FIGURE 8

The distance 5 is just half the length of R and it occurs to us to ask how far apart these segments AC and DE might be. This is found from the observation that the area of $\triangle ADC$ is one-quarter of S:

$$\triangle ADC = \frac{1}{2} AC \cdot \text{altitude } DF = \frac{1}{2} \cdot 5 \cdot DF = \frac{1}{4} \cdot 20 = 5,$$

yielding the very welcome result

$$DF = 2, \quad \text{the width of } R.$$

Thus R can be overlaid on S between the lines of AC and DE. With the right angles at the end of R, sliding R all the way down to A would put a triangular piece of R outside S; on the other hand, stopping at DF would leave the little triangle ADF still to be covered. Either way it's unsettled, but let's proceed by cutting a piece off the end of R to fit the less ambitious region $DFCE$. Since FC is not as long as DE, and our next step is to see how long it is:

in right triangle ADF, we have $AD = \sqrt{5}$ and $DF = 2$; hence by the Pythagorean theorem, $AF = 1$, making $FC = 4$.

Thus a possible first cut across R is one of length $\sqrt{5}$ from the midpoint Q of side PL to a point V which is four units from the corner U (Figure 9).

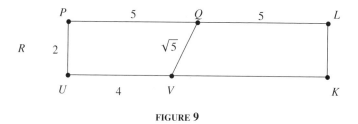

FIGURE 9

The remainder of S can be viewed as three triangles—the congruent pair $\triangle ABC$ and $\triangle DGE$, and $\triangle ADF$. We have all the dimensions of these triangles:

$\triangle ADF$ has sides of lengths $\sqrt{5}$, 2, and 1;
$\triangle ABC$ and $\triangle DGE$ have sides of lengths 5, $2\sqrt{5}$, and $\sqrt{5}$.

The problem remaining, then, is how to fit these lengths into the region $QVKL$ of R in a way that yields such triangles.

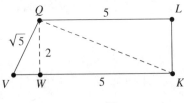

FIGURE 10

Our first observation is that a perpendicular QW to VK makes $\triangle QVW$ congruent to $\triangle ADF$, suggesting that the final cut be the diagonal of rectangle $QWKL$ (Figure 10). However, this is no good since the side of length 5 in the required triangles is the hypotenuse not a leg as in $\triangle QWK$.

Thankfully, the small section to cover $\triangle ADF$ can equally well be cut from the other end of R as $\triangle KLN$ by simply making $NK = 1$ and $LN = \sqrt{5}$ (Figure 11).

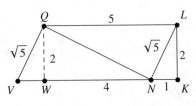

FIGURE 11

This leaves parallelogram $QVNL$, whose diagonal QN splits it into congruent triangles with the desired sides of lengths 5, $\sqrt{5}$, and $2\sqrt{5}$:

We have already noted that $QL = 5$ and $LN = \sqrt{5}$;
now, $NK = 1$, and $WK = QL = 5$, making $WN = 4$;
since $QW = 2$, the hypotenuse QN of $\triangle QWN$ is $\sqrt{20} = 2\sqrt{5}$,

as desired.

Finally, then, we obtain the desired dissection (Figure 12).

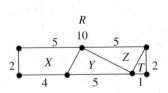 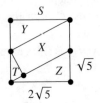

FIGURE 12

2. The Problem Contest Book V

The contests which are covered in this excellent book (prepared by George Berzsenyi and Stephen Maurer, Anneli Lax New Mathematical Library Series, volume 38, 1997) are the American High School Mathematics Examinations (AHSME) and the American Invitational Mathematics Examinations (AIME) for the years 1983–1988. A great many interesting problems and clever solutions are presented and the exposition is so well done that the book is a genuine pleasure to read.

The problems on these examinations are posed in multiple-choice form. We shall not retain this feature but simply enjoy them as straightforward challenges.

1. (AIME, 1988, #11)

An ellipse has foci at $(9, 20)$ and $(49, 55)$ in the xy-plane and is tangent to the x-axis. What is the length of its major axis?

The key to this problem is the reflector property of the ellipse: the focal radii to a point P on an ellipse make equal angles with the tangent at P (Figure 13).

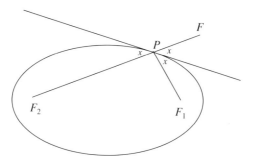

FIGURE 13

Thus, if focal radius $F_1 P$ is reflected in the tangent at P, its image is the equal segment PF which lies along the extension of $F_2 P$. It follows that $F_2 F$ is the sum of the focal radii, that is, the length of the major axis of the ellipse.

In the problem at hand then, the major axis is simply the length of the segment joining focus $(9, 20)$ and the image $(49, -55)$ of the reflection of the focus $(49, 55)$ in the x-axis:

$$\text{length of the major axis} = \sqrt{40^2 + (75)^2} = \sqrt{7225} = 85.$$

2. (AHSME, 1985, #30)

   Let $[x]$ denote the integer part of x, that is, the greatest integer $\leq x$. Determine the number of real solutions of the equation

   $$4x^2 - 40[x] + 51 = 0.$$

For real x, x^2 must be at least zero, and therefore

$$4x^2 = 40[x] - 51 \geq 0,$$

making

$$[x] \geq \frac{51}{40} > 1.$$

Thus $x \geq 2$.

Suppose x lies in the half-open unit interval $[k, k+1[$, k a positive integer ≥ 2, that is,

$$k \leq x < k+1.$$

Then $[x] = k$, and we have

$$4x^2 = 40[x] - 51 = 40k - 51,$$

and

$$x = \frac{1}{2}\sqrt{40k - 51}.$$

Since x lies in $[k, k+1[$, then

$$k \leq \frac{1}{2}\sqrt{40k - 51} < k+1,$$

$$k^2 \leq \frac{40k - 51}{4} < k^2 + 2k + 1,$$

$$4k^2 \leq 40k - 51 < 4k^2 + 8k + 4.$$

Thus k is restricted by the simultaneous inequalities

$$4k^2 - 40k + 51 \leq 0, \quad \text{i.e., } (2k - 3)(2k - 17) \leq 0,$$

and
$$4k^2 - 32k + 55 > 0, \quad \text{i.e., } (2k-5)(2k-11) > 0.$$

The first of these confines k to the interval $\left[\frac{3}{2}, \frac{17}{2}\right]$, i.e., to the integers 2, 3, 4, 5, 6, 7, 8, and the second requires k to be less than $\frac{5}{2}$ or more than $\frac{11}{2}$, i.e., to be in the infinite set $\{2, 6, 7, 8, 9, \ldots\}$ (recall $k \geq 2$). In satisfying these restrictions simultaneously, k can only be the four integers 2, 6, 7, 8, giving four solutions $x = \frac{1}{2}\sqrt{40k - 51}$ for these values.

3. The Leningrad Mathematical Olympiads 1987–1991

These problems come from the *Leningrad Mathematical Olympiads, 1987–1991*, compiled by Dimitry Fomin and Alexey Kirichenko (MathPro Press, 1994). We consider only three of these intriguing problems.

1. Squares *ABDE* and *ACFG* are drawn outwardly on sides *AB* and *AC* of $\triangle ABC$. If *EG* is parallel to *BC*, prove that $\triangle ABC$ is isosceles (Figure 14).

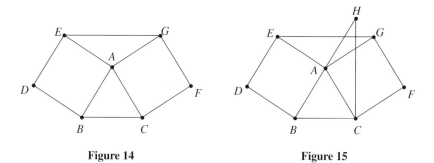

Figure 14   Figure 15

(Another problem concerning this configuration is given in part 6 of the essay Notes From *Quantum*.)

With squares around, it's always possible that a quarter-turn or a half-turn might be the key to things. The center of such a rotatin is often the center of a square, but this is compromised here by the presence of two squares. However, the squares are joined at the vertex *A*, and a clockwise quarter-turn of $\triangle AEG$ about *A* is the first step in a beautiful solution (Figure 15).

Clearly this rotation takes *G* to *C*, and since *EG* is parallel to *BC* to begin with, its image *HC* is perpendicular to *BC*. Also, $\angle BAH$, being composed of two right angles, is a straight angle. Furthermore, $AB = AE = AH$. Thus *A* is the midpoint of the hypotenuse of right triangle BCH and, as its circumcenter, is equidistant from the vertices. Hence $AB = AC$.

2. Let K be the point of intersection of the reflections of side BC in each of the other two sides of acute angled triangle ABC (Figure 16). Prove that AK passes through the circumcenter O of the triangle.

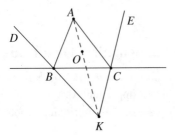

FIGURE 16

Clearly BA and CA are the bisectors of angles DBC and ECB, respectively. Hence A is equidistant from DK, BC, and EK, in particular from DK and EK, implying that A lies on the bisector of angle K.

Let $\angle DKA = \angle AKE = v$, and let $\angle BAK = u$ (Figure 17). Then for $\triangle ABK$, exterior $\angle DBA = u + v$, and since AB bisects $\angle DBC$, we have $\angle ABC$ is also equal to $u + v$. At B construct $\angle ABT = u$ to give T on AK. Then $\triangle ABT$ is isosceles with $AT = BT$. Moreover, since $\angle ABC = u + v$, then $\angle TBC = v$.

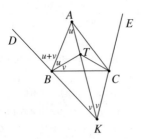

FIGURE 17

Thus TC subtends v at both B and K, implying $TBKC$ is cyclic. Thus, in circle $TBKC$, chords BT and TC are equal since they subtend the same angle v

at K on the circumference. Hence $TA = TB = TC$, and T is the circumcenter O of triangle ABC. The conclusion follows.

3. The greatest prime divisor in any of the prime decompositions of the n distinct positive integers $A_1, A_2, \ldots, A_n$ is the rth prime number p_r. Prove that, no matter how many numbers A_i there might be, the sum S of their reciprocals cannot be as great as p_r:

$$S = \frac{1}{A_1} + \frac{1}{A_2} + \cdots + \frac{1}{A_n} < p_r.$$

The prime decomposition of each A_i is $2^a 3^b \ldots p_r^t$. Hence $\frac{1}{A_i}$ is a term in the expansion of the product P of infinite geometric progressions

$$P = \left(1 + \frac{1}{2} + \frac{1}{2^2} + \cdots\right)\left(1 + \frac{1}{3} + \frac{1}{3^2} + \cdots\right) \cdots \left(1 + \frac{1}{p_r} + \frac{1}{p_r^2} + \cdots\right).$$

Thus

$$S < P = \frac{1}{1 - \frac{1}{2}} \cdot \frac{1}{1 - \frac{1}{3}} \cdots \cdots \frac{1}{1 - \frac{1}{p_r}},$$

giving

$$S < \frac{2}{1} \cdot \frac{3}{2} \cdot \frac{5}{4} \cdot \frac{7}{6} \cdot \frac{11}{10} \cdots \cdots \frac{p_{r-1}}{p_{r-1} - 1} \cdot \frac{p_r}{p_r - 1}. \qquad (1)$$

For $p_r = 2$, this yields $S < 2 = p_r$, and for $p_r = 3$, we get $S < 3 = p_r$. Now, for $p_r > 3$, consecutive prime numbers differ by at least 2, and we have

$$p_{r-1} < p_r - 1, \quad \text{implying} \quad \frac{p_{r-1}}{p_r - 1} < 1.$$

For $p_r > 3$, then, we have in line (1) above, upon cancelling the 2's and pushing the numerators forward one place, that

$$S < \left[\frac{3}{4} \cdot \frac{5}{6} \cdot \frac{7}{10} \cdots \cdots \frac{p_{r-1}}{p_r - 1}\right] \cdot p_r < p_r.$$

4. Problem-Solving Through Problems

Problem-Solving Through Problems is a beautifully written book by Loren Larson (Springer-Verlag, 1983) that has excitement on every page.

1. (Problem 3.3.24, page 106)

   Prove that, if the set of prime decompositions of $n + 1$ positive integers $a_1, a_2, \ldots, a_{n+1}$ contains powers of only n different prime numbers, then the product of some subset of the a's is a perfect square.

   It is a simple matter to determine the greatest square that divides an integer if you know its prime decomposition. For example,

   $$2 \cdot 3^2 \cdot 5^4 \cdot 7^3 \cdot 11 = (3^2 \cdot 5^4 \cdot 7^2) \cdot (2 \cdot 7 \cdot 11) = (3 \cdot 5^2 \cdot 7)^2 \cdot (2 \cdot 7 \cdot 11).$$

   A "square-free" factor $2 \cdot 7 \cdot 11$ is built up at the end by stripping a factor from each prime power that has an odd exponent. This converts the altered prime powers into squares, the product of which is the greatest square that divides the number. From the prime decomposition, then, it is easy to write the number a_r in the form $a_r = b_r^2 s_r$, where b_r^2 is the greatest square that divides a_r and s_r is the square-free part of a_r.

   Clearly a product

   $$a_i a_j \cdots a_k = (b_i^2 s_i)(b_j^2 s_j) \cdots (b_k^2 s_k)$$
   $$= (b_i b_j \cdots b_k)^2 \cdot s_i s_j \cdots s_k$$

   is a square if and only if the product of the square-free parts $s_i s_j \ldots s_k$ is a square. Evidently the name "square-free" is due to the fact that it is built up from prime factors that occur to the first degree. Now, it does no harm to include all the prime factors of the integer in its square-free part, provided each prime factor that really belongs there is given an exponent of 1 and the primes that don't are given an exponent of 0, as in

   $$2 \cdot 3^2 \cdot 5^4 \cdot 7^3 \cdot 11 = (3 \cdot 5^2 \cdot 7)^2 \cdot (2 \cdot 3^0 \cdot 5^0 \cdot 7 \cdot 11).$$

   If the prime decomposition of a_r is

   $$a_r = p_1^{t_1} p_2^{t_2} \cdots p_n^{t_n},$$

   where the primes are written in a fixed order (often in increasing order), then its square-free part may be specified by a 0–1 n-tuple $(u_1, u_2, \ldots, u_n)$ where u_j is 1 or 0 depending on whether the prime p_j is or is not in its makeup:

   thus the square-free part of $2 \cdot 3^2 \cdot 5^4 \cdot 7^3 \cdot 11$ is given by the vector $(1, 0, 0, 1, 1)$.

Let the square-free n-tuples of the given integers constitute the rows of a 0–1 matrix M (Figure 18); thus M has $n + 1$ rows, one for each integer a_i, and n columns, one for each prime p_j, and the entry in position ij is the exponent of the prime p_j in the square-free part of a_i.

M:

	p_1	p_2	$\cdots$	p_j	$\cdots$	p_n
a_1	0	1	$\cdots$	0	$\cdots$	0
a_2	1	1	$\cdots$	1	$\cdots$	1
$\vdots$						
a_i	1	0	$\cdots$	0	$\cdots$	1
$\vdots$						
a_{n+1}	0	0	$\cdots$	0	$\cdots$	0

FIGURE 18

Now, the product of the square-free parts of a subset of the a's is a square if and only if the **sum** of the corresponding rows of M contains only even components. Thus, we would like to show that some subset of the rows, when added as a (mod 2) sum, yields a row of 0's, i.e., the zero vector.

Since the rank of M over Z_2 cannot exceed n, the number of its columns, the $(n + 1)$ rows $r_1, r_2, \ldots, r_{n+1}$ must be linearly dependent. Thus, for some set of coefficients c_i, not all zero, there exists a (mod 2) sum

$$c_1 r_1 + c_2 r_2 + \cdots + c_{n+1} r_{n+1} = 0.$$

The c's which are equal to 1, of which there must be at least one since they are not all zero, identify a subset of the a's, the product of whose square-free parts, and hence of the a's themselves, is a perfect square, and our proof is complete.

2. (Problem 4.2.16(b), page 131: from the 1952 Putnam Examination)

Let $f(x) = a_n x^n + a_{n-1} x^{n-1} + \cdots + a_1 x + a_0$ be a polynomial of degree n with integer coefficients. Prove that, if a_0, a_n, and $f(1)$ are odd, then $f(x) = 0$ has no rational root.

Proceeding indirectly, suppose $f(x) = 0$ has a rational root $\frac{p}{q}$, where p and q are in their lowest terms. Then

$$a_n \left(\frac{p}{q}\right)^n + a_{n-1} \left(\frac{p}{q}\right)^{n-1} + \cdots + a_1 \left(\frac{p}{q}\right) + a_0 = 0$$

and

$$a_n p^n + a_{n-1} p^{n-1} q + \cdots + a_1 p q^{n-1} + a_0 q^n = 0. \qquad (A)$$

Since p divides the right side and is a factor of every term on the left side except $a_0 q^n$, p must also divide $a_0 q^n$, and since p and q are relatively prime, it follows that p divides a_0. Since a_0 is odd, then p must also be odd. Similarly, q divides $a_n p^n$, which implies q divides a_n, and we conclude that q is also odd.

It is given that

$$f(1) = a_n + a_{n-1} + a_{n-2} + \cdots + a_1 + a_0$$

is odd, implying that there is an odd number of odd coefficients.

From (A) above, however, we have that the left side,

$$a_n p^n + a_{n-1} p^{n-1} q + \cdots + a_1 p q^{n-1} + a_0 q^n,$$

is an even number. Now, with a_n, p, q, and a_0 all odd, all the products $p^s q^t$ are odd, and so each of these terms has the parity of its coefficient. Hence there is an odd number of odd terms, implying the contradiction that the sum must be odd. The conclusion follows.

SECTION 6
An Application of Turán's Theorem*

1. Suppose eight points are taken in the plane. In Figure 1 it is shown how 16 edges can be inserted between pairs of these points without filling in all three edges of any triangle that is determined by three of the points. However, no matter how one tries to put in 17 edges, some three of them will be found to close a triangle. In general, the insertion of more than $n^2/4$ edges in a set of n points (also called vertices), always forces a triangle to be formed. A further generalization of this result marked the birth of the subject *extremal graph theory* in a German forced-labor camp in 1940. The author was the Hungarian mathematician Paul Turán, who found that the contemplation of mathematics helped to sustain him during a most difficult period of his life.

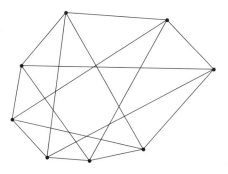

FIGURE 1

If all $\binom{n}{2}$ possible edges are present in a graph having n vertices, the graph is called a *complete graph* and is denoted by K_n. Now, one can insert only so

---

*This is a reworked version of material from my "Mathematical Gems" column in the *Two-Year College Mathematics Journal*, Vol. 11, June, 1980.

many edges in a set of n vertices before reaching the point of being unable to avoid filling in all $\binom{m}{2}$ edges determined by some m of the vertices. Turán's theorem specifies the maximum number of edges that can be inserted in a set of n vertices before the graph is forced to contain a K_m. The proof is too long and complicated for inclusion in a popular essay, but we can enjoy using the theorem.

2. As might be expected, a general formula for this maximum is somewhat involved. The expression is simplified a little if it is stated for a K_{m+1} instead of a K_m. Suppose dividing n by m gives a quotient q and remainder r, that is,

$$n = qm + r, \quad \text{where } 0 \le r < m.$$

Then, in terms of n, m, q and r, Turán's theorem may be stated as follows.

Turán's Theorem. *If a graph with n vertices fails to contain a K_{m+1}, then it cannot contain more edges than*

$$\tfrac{1}{2}r(q+1)[n-(q+1)] + \tfrac{1}{2}q(m-r)(n-q).$$

For $n = 8$ and $K_{m+1} = K_3$ for triangles, we have $m = 2$, and since $8 = 4 \cdot 2 + 0$, then $q = 4$ and $r = 0$, and the formula gives the maximum

$$0 + \tfrac{1}{2} \cdot 4(2-0)(8-4) = 16,$$

as noted above.

Now let us apply this remarkable result to a neat problem of Paul Erdős, a fellow Hungarian and pioneer in extremal graph theory and extremal geometry.

3. The problem concerns sets S of points in the plane with the property that no two of the points determine a distance exceeding 1. This condition is described by saying that S has *diameter* ≤ 1 (the diameter of a set is the least upper bound of the distances determined by pairs of its points). Clearly the points of a set of diameter ≤ 1 cannot be spread very far apart and a certain amount of crowding is inevitable as the number of points is increased. However, there are times when we would like to arrange the points so as to produce as many "big" distances as possible. The question therefore arises as to the maximum number of distances which can be made to exceed a stipulated magnitude d. We shall solve this difficult problem only in the case of $d = 1/\sqrt{2}$, which is approximately .707.

An Application of Turán's Theorem

Consider, for example, a set of six points. Placed at the vertices of a regular hexagon which has main diagonal 1 (Figure 2), it is easy to calculate that the edges of the hexagon are too small (at $\frac{1}{2}$), but that the six minor diagonals qualify (at $\frac{\sqrt{3}}{2}$), and together with the three major diagonals of unit length, yield a total of nine distances exceeding $\frac{1}{\sqrt{2}}$.

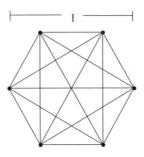

FIGURE 2

This is a fair crop of big distances and it is not immediately obvious how to do any better. However, the number can be raised to 12 by placing three of the points at the vertices of an equilateral triangle of side 1 and putting the remaining three points inside the triangle, one near each vertex as in Figure 3. Of the fifteen distances they determine, the only small ones are the three from a vertex to its nearby point.

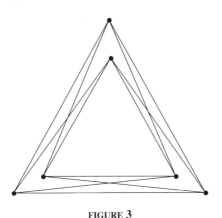

FIGURE 3

It turns out that this latter scheme can be generalized to a set of n points, and with the aid of Turán's theorem we shall prove the following engaging theorem of Erdős, where $[x]$ denotes the greatest integer $\leq x$.

Theorem (Erdős). *If S is a set of n points in the plane having diameter ≤ 1, the number of distances $> \frac{1}{\sqrt{2}}$ that are determined by S cannot exceed $[n^2/3]$; moreover, for every n, there exists a set of n such points which attains the maximum $[n^2/3]$.*

Let S denote any set of n points in the plane with diameter ≤ 1 and let each pair of points which determines a distance exceeding $1/\sqrt{2}$, and only those pairs, be joined by an edge. Thus a graph G is produced having n vertices and an edge for each distance exceeding $1/\sqrt{2}$. Let us consider the possibility of G containing a K_4.

There are only three ways in which four vertices can be arranged in the plane (Figure 4):

(i) 3 or 4 on a line (ii) one inside the triangle formed by the other 3 (iii) a convex quadrilateral

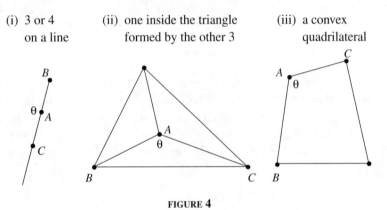

FIGURE 4

The four vertices of a K_4 in G would have to assume one of these arrangements, and it is evident that, in every case, some two adjacent edges AB and AC of the K_4 would have to be inclined at an angle $\theta = \angle BAC$ which is at least a right angle. The length of the side BC in such a triangle ABC is given by the law of cosines:

$$BC^2 = AB^2 + AC^2 - 2AB \cdot AC \cdot \cos\theta.$$

Now, each of AB and AC exceeds $1/\sqrt{2}$, making their squares exceed $\frac{1}{2}$, and since θ is at least a right angle, then $\cos\theta \leq 0$. Hence

$$BC^2 > \tfrac{1}{2} + \tfrac{1}{2} + 0 = 1,$$

contradicting the assumption that $BC \leq 1$. It follows that the graph G cannnot contain a K_4, that is, a K_{3+1}. By Turán's theorem, then, the maximum number of edges in G is

$$N = \tfrac{1}{2}r(q+1)[n-(q+1)] + \tfrac{1}{2}q(m-r)(n-q),$$

where

$$n = 3q + r, \quad \text{and} \quad 0 \le r < 3.$$

It remains only to go through the simple calculations that show $N = [n^2/3]$.

Noting that $m = 3$, multiplying out and simplifying, we easily obtain

$$N = 3q^2 + 2qr + \frac{r(r-1)}{2},$$

while

$$\frac{n^2}{3} = 3q^2 + 2qr + \frac{r^2}{3}.$$

Now, r is either 0, 1, or 2. For $r = 0$ or 1, we have $r^2/3 < 1$, and we get

$$[n^2/3] = 3q^2 + 2qr,$$

which is also the value of N since, in each case, the fraction $r(r-1)/2$ vanishes. If $r = 2$, then

$$\left[\frac{n^2}{3}\right] = \left[3q^2 + 2qr + \frac{4}{3}\right] = 3q^2 + 2qr + 1,$$

which is again the value of N since $r(r-1)/2 = 1$ in this case.

It remains to construct a set of n points of diameter ≤ 1 which has exactly $[n^2/3]$ distances $> 1/\sqrt{2}$.

As in the case of six points, we begin by placing three of the points at the vertices of an equilateral triangle ABC of side 1. Again we wish to arrange the remaining points inside $\triangle ABC$ in little clusters near the vertices. Let us choose a small radius r and distribute the points as equally as possible in the interiors of circular sectors having centers at A, B, and C, and radius r (Figure 5). If r is chosen so that

$$1 - 2r > \frac{1}{\sqrt{2}}, \quad \text{that is,} \quad r < \frac{1}{2}\left(1 - \frac{1}{\sqrt{2}}\right),$$

then it is not difficult to see that the distance between two points in different clusters will exceed $1/\sqrt{2}$:

Let points D and E on AB be such that $AD = BE = r$ (Figure 5). Then $DE = 1 - 2r > 1/\sqrt{2}$. Now let DF and EG be perpendicular to AB. Then

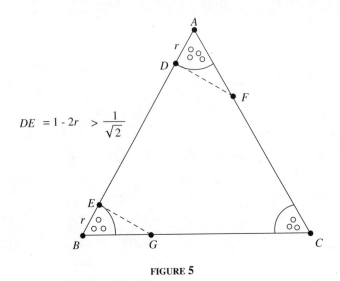

FIGURE 5

DF and *EG* are tangents to the sectors at *A* and *B*, respectively, and the distance between a point in *A*'s cluster and a point in *B*'s cluster must exceed the perpendicular distance *DE* between these parallel tangents.

Now, n is either of the form $3k$, $3k+1$, or $3k+2$. Suppose, for example, that $n = 3k+1$. Then $n^2 = 9k^2 + 6k + 1$ and

$$\left[\frac{n^2}{3}\right] = \left[3k^3 + 2k + \frac{1}{3}\right] = 3k^2 + 2k.$$

Also, the clusters will contain, respectively, k, k, and $(k+1)$ points, counting the vertices A, B, and C. Consequently, the number of distances between clusters, that is, the number of distances $> 1/\sqrt{2}$, will be exactly

$$k(k) + k(k+1) + k(k+1) = 3k^2 + 2k = \left[\frac{n^2}{3}\right].$$

The other cases are handled similarly.

Finally, you might enjoy the following exercise.

Fourteen forest rangers patrol a circular park having diameter nine miles. Each ranger is equipped with a walkie-talkie which has a range of seven miles. Prove that, however they might move about the park, at any given time, one of the rangers will be able to talk to at least four of the others, and that, if this ranger is unable to reach all 13 of the others, then there will be a second ranger who can communicate with four other rangers.

A full proof of Turán's theorem and a detailed discussion of its application to the above theorem of Erdős is given by Bondy and Murty in their excellent text *Graph Theory with Applications* (American Elsevier, 1976, pages 113–115).

SECTION 7
Four Problems from Putnam Papers

1. (1947)

    Prove that, if a, b, c, d are distinct integers, then the only possible integral root of the equation

    $$(x-a)(x-b)(x-c)(x-d) - 4 = 0$$

    is

    $$x = \frac{a+b+c+d}{4}.$$

    If x is an integer, then so is each factor $x-a, x-b, x-c, x-d$, and if their product is 4, their magnitudes in some order must be $(1, 1, 1, 4)$ or $(1, 1, 2, 2)$. Now, in any set of three numbers, some two have the same sign, and so if three of the factors have the same magnitude, two of them would be identical, implying the contradiction that a, b, c, d are not distinct. Hence the magnitudes must be $(1, 1, 2, 2)$.

    Similarly, in order to avoid two equal values among a, b, c, d, two factors with the same magnitude must have opposite signs. Thus the factors can only be ± 1 and ± 2, with sum zero, in which case

    $$(x-a) + (x-b) + (x-c) + (x-d) = 4x - (a+b+c+d) = 0$$

    and

    $$x = \frac{a+b+c+d}{4}.$$

2. (1969)

    Observe that 96 is a multiple of 24 and that the sum of all the positive divisors of 95 is $1 + 5 + 19 + 95 = 120$, which is also a multiple of 24.

    Prove that whenever a positive integer $n + 1$ is a multiple of 24, so is the sum of all the positive divisors of n.

If $n+1$ is a multiple of 24, then, for some positive integer k, $n = 24k - 1$. Suppose a and b is a pair of complementary divisors of n, that is,

$$ab = n = 24k - 1.$$

Then neither a nor b can be divisible by 2 or 3.

Now, while we have a little information about ab, it is $a + b$ that figures in the problem, and so we need to look around for a way of relating these quantities. It is pretty obvious that multiplying $a + b$ by a or b brings ab into play. Consider, then, the product

$$\begin{aligned} a(a+b) &= a^2 + ab \\ &= a^2 + 24k - 1 \\ &= a^2 - 1 + 24k \\ &= (a-1)(a+1) + 24k. \end{aligned}$$

Since a is odd, $a - 1$ and $a + 1$ are consecutive even numbers, and so, while each is divisible by 2, one is divisible by 4, implying $(a-1)(a+1)$ is divisible by 8. Also, one of the three consecutive integers $a-1, a, a+1$ is divisible by 3, and since it isn't a, it must be either $a-1$ or $a+1$, implying that $(a-1)(a+1)$ must be divisible by 3. Because 8 and 3 are relatively prime, then $(a-1)(a+1)$ is divisible by 24, and it follows that $a(a+b)$ is also divisible by 24.

But a has no factor in common with $24 (= 2^3 \cdot 3)$, and so $a(a+b)$'s factor 24 must come entirely from the factor $(a+b)$. That is to say, the sum of each pair of complementary divisors of n is divisible by 24. The desired conclusion follows, then, if we can show that all the divisors of n go together into complementary pairs.

Since each divisor $< \sqrt{n}$ pairs with one $> \sqrt{n}$, the complementary pairs will comprise all the divisors unless n is a perfect square, in which case there is an extra self-complementary divisor $\sqrt{n}$. Suppose, then, that $n = a^2$. Since n is odd, so is a. Thus the proof is completed with the simple observation that

$$a^2 = 24k - 1$$

gives

$$a^2 - 1 = 24k - 2,$$

which is an impossible equation since, as we have seen, the left side is divisible by 8, while clearly the right side isn't.

3. (1996)

Find the least number A such that, for any two squares, the sum of whose areas is 1, a rectangle of area A exists into which the squares can be packed without overlapping of interior points. (Assume that the squares are to be packed with their sides parallel to the sides of the rectangle.)

Clearly the tightest packing is obtained for abutting squares (Figure 1). Since the sides of the squares are numbers the sum of whose squares is unity, they are the sine and cosine of some angle $x > 0$. Let us assign the labels so that $\cos x \geq \sin x$, thus restricting x to the half closed interval $]0, \frac{\pi}{4}]$. Hence the smallest rectangle into which the squares can be packed has area

$$M = (\cos x + \sin x)\cos x = \cos^2 x + \sin x \cos x = \cos^2 x + \tfrac{1}{2}\sin 2x.$$

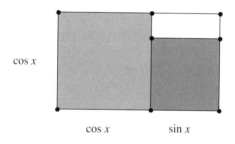

FIGURE 1

In order to accommodate all such pairs of squares, the required area A is the maximum value of M as x varies in $]0, \frac{\pi}{4}]$. Differentiating, we get

$$\frac{dM}{dx} = -\sin 2x + \cos 2x,$$

which takes the value zero for $2x = \pi/4$. Since

$$\frac{d^2 M}{dx^2} = -2\cos 2x - 2\sin 2x < 0,$$

M is indeed a maximum at $x = \pi/8$.
Now,

$$\frac{1}{\sqrt{2}} = \cos\frac{\pi}{4} = 2\cos^2\frac{\pi}{8} - 1, \quad \text{and so} \quad \cos^2\frac{\pi}{8} = \frac{\sqrt{2}+1}{2\sqrt{2}},$$

giving

$$A = \cos^2 \frac{\pi}{8} + \frac{1}{2}\sin\frac{\pi}{4} = \frac{\sqrt{2}+2}{2\sqrt{2}} = \frac{1+\sqrt{2}}{2}.$$

4. (1996)

Let C_1 and C_2 be circles whose centers are 10 units apart and whose radii are 1 and 3. Find, with proof, the locus of all points M for which there exist points X on C_1 and Y on C_2 such that M is the midpoint of XY.

Let the centers of C_1 and C_2 be O_1 and O_2, respectively, and let the line through the centers meet C_1 at P and Q (Figure 2). Now, for a fixed point X on C_1, the locus of M as Y goes around C_2 is the image circle C_2' of C_2 under the dilatation $X(\frac{1}{2})$. The radius of C_2' is one-half that of C_2, that is, $\frac{3}{2}$, and its center O_2' is the image of O_2 under the dilatation, namely the midpoint of XO_2.

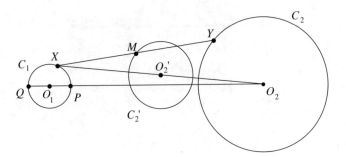

FIGURE 2

To each point X on C_1 there corresponds a center O_2' and a circle C_2' of points on the locus of M. Now, these centers are the midpoints of segments from O_2 to a point X on C_1, and so altogether they constitute the image circle C_1' of C_1 under the dilatation $O_2(\frac{1}{2})$ (Figure 3).

Clearly, C_1' has its center O on $O_1 O_2$ and it crosses $O_1 O_2$ at the midpoints A and B of $O_2 P$ and $O_2 Q$ (Figure 3). Since $O_1 O_2 = 10$ and the radius of C_1 is 1, then

$$O_2 A = \frac{9}{2}, \quad O_2 B = \frac{11}{2}, \quad \text{making } AB = 1.$$

Thus $O_2 O = O_2 A + AO = \frac{9}{2} + \frac{1}{2} = 5$, and the center O of C_1' is in fact the midpoint of $O_1 O_2$, and its radius is $\frac{1}{2}$.

Four Problems from Putnam Papers

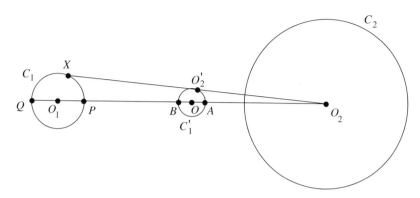

FIGURE 3

Thus the locus of M is the region traced out by the circumference of C_2', a circle of radius $\frac{3}{2}$, as its center traverses C_1'. It is clear that when the center of C_2' is at A, the distances from O of the points on its circumference run from a minimum of one unit at S to a maximum of two units at T (Figure 4). Thus as C_2' goes around C_1', the symmetrical locus of M is an annulus K centered at O with inner radius 1 and outer radius 2.

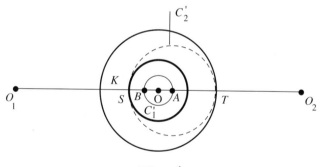

FIGURE 4

SECTION **8**
Topics Based on Problems from *Quantum*

Although it breaks my heart to say it, *Quantum* is no longer being published as of 2002. *Quantum* is a magazine about Mathematics and Science that is directed especially at secondary school and undergraduate university students. During its tenure it was published every two months by the National Science Teachers Association in cooperation with the Quantum Bureau of the Russian Academy of Sciences. Since its inception in 1990, its pages have been filled with articles of highest quality on subjects of popular interest. I don't know of any other magazine that makes mathematics and science so attractive to students. We can only hope that it will be revived in the near future.

1. (Math Investigations: Geometry in the Pagoda, Problem 2, page 48, Jan./Feb., 1995)

    Equal circles, O_1 and O_2, are tangent to each other and to the sides of square $ABCD$ (Figure 1). The tangent from A to O_1 meets the tangent from D to O_2 at E. Prove that the incircle O_3 of $\triangle AED$ is the same size as O_1 and O_2.

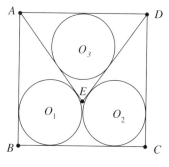

FIGURE 1

Let the length of a side of the square be $4r$. Then the radius of each of O_1 and O_2 is r. Letting the radius of O_3 be t, we want to show $t = r$ (Figure 2).

49

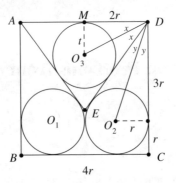

FIGURE 2

Clearly, DO_3 and DO_2 bisect angles ADE and EDC, and this tells us that $2x + 2y = 90°$ and $x + y = 45°$. Now, clearly

$$\tan y = \frac{1}{3},$$

and since O_3 touches AD at its midpoint M,

$$\tan x = \frac{t}{2r}.$$

Hence

$$1 = \tan 45° = \tan(x + y) = \frac{\tan x + \tan y}{1 - \tan x \cdot \tan y},$$

$$\tan x + \tan y = 1 - \tan x \cdot \tan y,$$

that is,

$$\frac{t}{2r} + \frac{1}{3} = 1 - \frac{t}{6r},$$
$$3t + 2r = 6r - t,$$

and the desired

$$t = r.$$

2. (Challenges in Physics and Math M171, May/June, 1996, page 31: proposed by A. Perlin)

Does there exist a quadratic polynomial $f(x)$ with integral coefficients with the unusual property that, whenever x is a positive integer which consists only of 1's, $f(x)$ is also a positive integer consisting only of 1's.

A positive integer x which consists entirely of 1's is of the form $\frac{1}{9}(10^k-1)$. Thus we wish to determine coefficients a, b, c such that, for all positive integers n, the expression

$$a\left(\frac{10^n-1}{9}\right)^2 + b\left(\frac{10^n-1}{9}\right) + c$$

is also a number of this form $\frac{1}{9}(10^k - 1)$. Now,

$$a\left(\frac{10^n-1}{9}\right)^2 + b\left(\frac{10^n-1}{9}\right) + c$$
$$= \frac{1}{9}\left[\frac{a}{9}(10^{2n} - 2\cdot 10^n + 1) + b(10^n - 1) + 9c\right]$$
$$= \frac{1}{9}\left[\frac{a}{9}\cdot 10^{2n} + \left(b - \frac{2a}{9}\right)10^n + \left(\frac{a}{9} - b + 9c\right)\right].$$

This would be just $\frac{1}{9}(10^{2n} - 1)$ if $\frac{a}{9} = 1$, $b - \frac{2a}{9} = 0$, and $\frac{a}{9} - b + 9c = -1$, that is, if

$$a = 9, \quad b = 2, \quad \text{and} \quad c = 0.$$

Hence a suitable polynomial is

$$f(x) = 9x^2 + 2x.$$

It is easy to confirm that this polynomial $f(x) = x(9x + 2)$ does have the property in question:

if $x = 11\ldots 1$, containing m 1's,
then $9x + 2 = 99\ldots 9 + 2 = 100\ldots 01 = 10^m + 1$,
and

$$f(x) = 11\ldots 1(10^m + 1)$$
$$= 11\ldots 100\ldots 0 + 11\ldots 1$$
$$= 11\ldots 111\ldots 1, \text{ containing } 2m \text{ 1's.}$$

3. (The Schwab–Schoenberg Mean)

Encountering the beautiful article "Remarkable Limits" by M. Crane and A. Nudelman in the July/Aug 1997 issue of *Quantum*, I was moved to dig out some notes I made on the subject back in the 1970s. Schwab's work is so striking and Schoenberg's alternative approach so elegant that I can't resist including an account of them in this collection of essays.

I. The Result

The subject is a pair of interlocking sequences $\{a_n\}$ and $\{b_n\}$ defined as follows:

Let a and b be positive numbers with $a < b$.

Let $a_0 = a$, $b_0 = b$, and for $n \geq 0$,

$$a_{n+1} = \frac{a_n + b_n}{2}, \quad b_{n+1} = \sqrt{a_{n+1}b_n}.$$

Thus a_{n+1} is just the arithmetic mean of a_n and b_n, while b_{n+1} is the geometric mean, not of a_n and b_n as you might expect, but of a_{n+1} and b_n. Thus a_{n+1} must be calculated before b_{n+1} can be determined. For example, for $a = 1$ and $b = 2$, we have

n	0	1	2	3	4	—
a_n	1	1.5	1.61602...	1.64452...	1.65165...	—
b_n	2	1.73205...	1.67302...	1.65871...	1.65515...	—

It appears that $\{a_n\}$ is increasing and $\{b_n\}$ decreasing and that they might be approaching a common limit. These sequences do, in fact, approach the limit 1.6539867.... The intriguing general result concerning such sequences was established geometrically by Schwab in 1813 and rediscovered by Borchardt in 1880:

$\{a_n\}$ and $\{b_n\}$ approach the common limit $\dfrac{\sqrt{b^2 - a^2}}{\arccos \frac{a}{b}}$.

Nowadays this is called the Schwab–Schoenberg mean since, in more recent times, Isaac Schoenberg has given a nice analytic proof.

II. Schwab's Proof

(a) Before launching into Schwab's wonderful discovery we need to establish that $\{a_n\}$ and $\{b_n\}$ do approach a common limit. It all follows nicely from the fact that, since b is greater than a to begin with, b_n is greater than a_n all down the line. A simple induction is sufficient to show this:

Since $a < b$, then $a_0 < b_0$. Suppose, for some $n \geq 0$, that $a_n < b_n$. Since the arithmetic mean a_{n+1} is halfway between a_n and b_n, we

have
$$a_n < a_{n+1} < b_n.$$
Thus
$$a_{n+1} = \sqrt{a_{n+1}a_{n+1}} < \sqrt{a_{n+1} \cdot b_n} = b_{n+1}.$$

Corollaries:
1. $a_n < a_{n+1}$, implying $\{a_n\}$ is increasing;
2. $b_{n+1} = \sqrt{a_{n+1}b_n} < \sqrt{b_n \cdot b_n} = b_n$, implying $\{b_n\}$ is decreasing;
3. $a_n < b_n \leq b$, implying $\{a_n\}$ is bounded above, and $b_n > a_n \geq a$, implying $\{b_n\}$ is bounded below.

It follows that $\{a_n\}$ approaches some limit α and $\{b_n\}$ approaches a limit β. Finally, from $a_{n+1} = (a_n + b_n)/2$, we obtain
$$\lim a_{n+1} = \frac{\lim a_n + \lim b_n}{2},$$
i.e.,
$$\alpha = \frac{\alpha + \beta}{2},$$
giving
$$\alpha = \beta.$$

(b) At a point A_0 on straight line XY erect perpendicular A_0O of length a_0 (Figure 3). With center O construct a circular arc of radius b_0 to meet XY at B_0 and B_0'. Let $\angle B_0OB_0' = 2\theta$. Since $\triangle B_0OB_0'$ is isosceles and OA_0

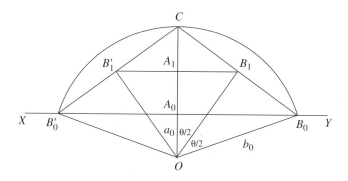

FIGURE 3

is perpendicular to the base, it follows that $\angle B_0OA_0 = \theta$. Let OA_0 meet the arc at C and let B_1 and B_1' be the midpoints, respectively, of CB_0 and CB_0'. Then B_1B_1' is parallel to B_0B_0' and one-half as long; moreover, B_1B_1' lies halfway between B_0B_0' and C and crosses A_0C perpendicularly at its midpoint A_1.

Hence

$$OA_1 = OC - CA_1$$
$$= OC - \frac{1}{2}CA_0$$
$$= b_0 - \frac{1}{2}(b_0 - a_0)$$
$$= \frac{a_0 + b_0}{2}$$
$$= a_1.$$

Since $\triangle OCB_0$ is isosceles and B_1 is the midpoint of the base B_0C, OB_1 is perpendicular to B_0C and bisects the vertical angle at O. Thus, in right triangle OCB_1, B_1A_1 is the altitude to the hypotenuse and we have the standard result

$$OB_1^2 = OC \cdot OA_1,$$

giving

$$OB_1 = \sqrt{b_0 \cdot a_1} = b_1.$$

Thus $\triangle OB_1B_1'$ is analogous to the original triangle OB_0B_0', being isosceles with arms equal to b_1 and altitude to the base equal to a_1, except that its base and vertical angle are each only one-half their predecessors:

$$B_1B_1' = \frac{1}{2}B_0B_0', \quad \text{and} \quad \angle B_1OB_1' = \frac{1}{2}\angle B_0OB_0' = \frac{2\theta}{2}.$$

Treating $\triangle B_1OB_1'$ as we treated $\triangle B_0OB_0'$, we obtain an isosceles triangle B_2OB_2' with

$$\text{arm } OB_2 = b_2, \quad \text{altitude } OA_2 = a_2,$$
$$\text{base } B_2B_2' = \frac{1}{2}B_1B_1' = \frac{1}{2^2}B_0B_0',$$
$$\text{and vertical angle } B_2OB_2' = \frac{1}{2}\angle B_1OB_1' = \frac{1}{2^2}\angle B_0OB_0' = \frac{2\theta}{2^2}.$$

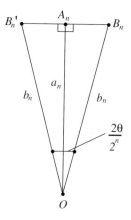

FIGURE 4

Continuing like this, the nth isosceles triangle $B_n OB'_n$ (Figure 4) would have

$$\text{arm } OB_n = b_n, \quad \text{altitude } OA_n = a_n,$$

$$\text{base } B_n B'_n = \frac{1}{2^n} B_0 B'_0,$$

$$\text{and vertical angle } B_n OB'_n = \frac{1}{2^n} \angle B_0 OB'_0 = \frac{2\theta}{2^n}.$$

It follows that a fan of 2^n copies of $\triangle B_n OB'_n$, around a circular arc of radius b_n, would spread over a central angle of

$$2^n \cdot \frac{2\theta}{2^n} = 2\theta$$

(Figure 5).

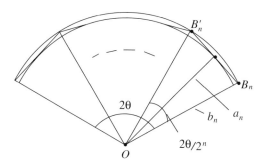

FIGURE 5

In this arc, the bases of the triangles determine an inscribed polygonal path P of length $2^n \cdot \frac{1}{2^n} B_0 B_0' = B_0 B_0'$. Also, the bases are tangent to a concentric arc of smaller radius a_n and thus circumscribe a path around it. While these arcs and the path P vary with n, the length of the polygonal path remains constant at $B_0 B_0'$.

From the definition arc length/radius = the central angle, the lengths of the arcs are simply $2\theta \cdot b_n$ and $2\theta \cdot a_n$. Since P is inscribed in one of the arcs and circumscribed about the other, its length lies between the lengths of the arcs, and for all n we have

$$2\theta \cdot a_n < B_0 B_0' < 2\theta \cdot b_n,$$

giving

$$a_n < \frac{B_0 B_0'}{2\theta} < b_n. \qquad (1)$$

From Figure 3 it is clear that

$$B_0 B_0' = 2 B_0 A_0 = 2\sqrt{b_0^2 - a_0^2} = 2\sqrt{b^2 - a^2}$$

and that $\cos\theta = (a_0/b_0) = (a/b)$. In the limit, then, since $\lim a_n = \lim b_n (= \alpha)$, (1) yields Schwab's marvelous result

$$\alpha = \frac{B_0 B_0'}{2\theta} = \frac{2\sqrt{b^2 - a^2}}{2\theta} = \frac{\sqrt{b^2 - a^2}}{\arccos \frac{a}{b}}.$$

III. Schoenberg's Proof

Since it has already been established in our account of Schwab's work that $\{a_n\}$ and $\{b_n\}$ converge to a common limit α, let us proceed with Schoenberg's argument from this point onward. He first directs our attention to the sequence $\{t_n\}$ where

$$t_n = \frac{\sqrt{b_n^2 - a_n^2}}{\arccos \frac{a_n}{b_n}}.$$

(The drawback with this approach is that one needs to know the final formula in advance.) It is not difficult to show the surprising result that all the terms in $\{t_n\}$ are the same. To this end, we have

$$t_{n+1} = \frac{\sqrt{b_{n+1}^2 - a_{n+1}^2}}{\arccos \frac{a_{n+1}}{b_{n+1}}} = \frac{\sqrt{a_{n+1} \cdot b_n - a_{n+1}^2}}{\arccos \frac{a_{n+1}}{\sqrt{a_{n+1} \cdot b_n}}} \quad (\text{recall } b_{n+1} = \sqrt{a_{n+1} \cdot b_n})$$

$$= \frac{\sqrt{a_{n+1}(b_n - a_{n+1})}}{\arccos \sqrt{\frac{a_{n+1}}{b_n}}}.$$

Now, $a_{n+1} = (a_n + b_n)/2$, and so

$$a_{n+1}(b_n - a_{n+1}) = \frac{a_n + b_n}{2}\left(b_n - \frac{a_n + b_n}{2}\right)$$
$$= \frac{b_n + a_n}{2}\left(\frac{b_n - a_n}{2}\right) = \frac{b_n^2 - a_n^2}{4}.$$

Next, let

$$\arccos\sqrt{\frac{a_{n+1}}{b_n}} = \theta, \quad \text{i.e., } \cos\theta = \sqrt{\frac{a_{n+1}}{b_n}}.$$

Then

$$\cos 2\theta = 2\cos^2\theta - 1$$
$$= \frac{2a_{n+1}}{b_n} - 1$$
$$= \frac{a_n + b_n}{b_n} - 1 \quad \left(\text{recall } a_{n+1} = \frac{a_n + b_n}{2}\right)$$
$$= \frac{a_n}{b_n}.$$

Hence

$$\arccos\frac{a_n}{b_n} = 2\theta = 2\arccos\sqrt{\frac{a_{n+1}}{b_n}}.$$

Thus

$$\arccos\sqrt{\frac{a_{n+1}}{b_n}} = \frac{1}{2}\arccos\frac{a_n}{b_n},$$

and

$$t_{n+1} = \frac{\sqrt{a_{n+1}(b_n - a_{n+1})}}{\arccos\sqrt{\frac{a_{n+1}}{b_n}}} = \frac{\frac{1}{2}\cdot\sqrt{b_n^2 - a_n^2}}{\frac{1}{2}\cdot\arccos\frac{a_n}{b_n}} = t_n.$$

Hence $\{t_n\}$ is indeed constant and

$$\lim t_n = t_0 = \frac{\sqrt{b_0^2 - a_0^2}}{\arccos\frac{a_0}{b_0}} = \frac{\sqrt{b^2 - a^2}}{\arccos\frac{a}{b}}.$$

It remains, then, only to show that $\lim t_n$ is in fact the common limit α of $\{a_n\}$ and $\{b_n\}$.

By definition,
$$\lim t_n = \lim \frac{\sqrt{b_n^2 - a_n^2}}{\arccos \frac{a_n}{b_n}}.$$

Now let
$$\arccos \frac{a_n}{b_n} = \varphi_n, \quad \text{where } 0 < \varphi_n < \frac{\pi}{2}.$$

Then $\cos \varphi_n = a_n/b_n$, and
$$\sin^2 \varphi_n = 1 - \cos^2 \varphi_n = 1 - \frac{a_n^2}{b_n^2} = \frac{b_n^2 - a_n^2}{b_n^2}.$$

Hence
$$\sqrt{b_n^2 - a_n^2} = \sqrt{b_n^2 \cdot \sin^2 \varphi_n} = b_n \sin \varphi_n, \quad \text{(recall } \varphi_n \text{ is acute)}.$$

Thus
$$\lim t_n = \lim \frac{\sqrt{b_n^2 - a_n^2}}{\arccos \frac{a_n}{b_n}}$$
$$= \lim \frac{b_n \sin \varphi_n}{\varphi_n}.$$

Since a_n and b_n approach the same limit, then $\cos \varphi_n = a_n/b_n \to 1$ and $\varphi_n \to 0$ from above. Hence
$$\lim t_n = (\lim b_n) \left(\lim \frac{\sin \varphi_n}{\varphi_n} \right) = \alpha \cdot 1 = \alpha.$$

IV. Remarks

It is exceedingly difficult to determine the common limit of the more symmetrically defined sequences
$$a_0 = a, b_0 = b, a < b, \quad \text{and}$$
$$a_{n+1} = \frac{a_n + b_n}{2} \quad \text{and} \quad b_{n+1} = \sqrt{a_n b_n}.$$

These sequences were mentioned by Lagrange in 1785 and came to the notice of the 14-year old Gauss in 1791. In 1799 Gauss dealt successfully with the case of $a = 1, b = \sqrt{2}$, but it wasn't until 1818 that he succeeded, with advanced techniques, in establishing the general result
$$\alpha = \left(\frac{2}{\pi} \int_0^{\pi/2} \frac{dx}{\sqrt{a^2 \sin^2 x + b^2 \cos^2 x}} \right)^{-1}.$$

For more on this subject of algorithms involving arithmetic and geometric means, you might enjoy the paper of the same name by B. C. Carlson in the *American Mathematical Monthly*, Vol. 78, 1971, beginning on page 496.

Exercise Use Schoenberg's approach to establish Carlson's discovery that

if $a_0 = a$, $b_0 = b$, $a < b$, and

$$a_{n+1} = \sqrt{\frac{a_n + b_n}{2} \cdot a_n}, \quad b_{n+1} = \sqrt{\frac{a_n + b_n}{2} \cdot b_n},$$

then $\alpha = \beta = \sqrt{\dfrac{b^2 - a^2}{2 \log \frac{b}{a}}}.$

4. (Problem M215, Sept/Oct 1997, page 27, proposed by V. Protasov)

Let $b(n)$ denote the number of ways of representing the nonnegative integer n in the form

$$n = a_0 + a_1 \cdot 2 + a_2 \cdot 2^2 + a_3 \cdot 2^3 + \cdots + a_k \cdot 2^k,$$

where the coefficients a_i are each either 0, 1, or 2. Calculate $b(1997)$.

Theoretically, nothing could be easier. Noting that $2^{11} = 2048 > 1997$, it is clear that $b(1997)$ is the coefficient of x^{1997} in the product

$$(x^{0 \cdot 2^0} + x^{1 \cdot 2^0} + x^{2 \cdot 2^0}) \cdot (x^{0 \cdot 2^1} + x^{1 \cdot 2^1} + x^{2 \cdot 2^1}) \cdot (x^{0 \cdot 2^2} + x^{1 \cdot 2^2} + x^{2 \cdot 2^2}) \cdots$$
$$\cdot (x^{0 \cdot 2^9} + x^{1 \cdot 2^9} + x^{2 \cdot 2^9})(x^{0 \cdot 2^{10}} + x^{1 \cdot 2^{10}} + x^{2 \cdot 2^{10}});$$

in each term in x^{1997}, the exponents of the terms chosen from these factors are in turn a set of terms $a_0 \cdot 2^0, a_1 \cdot 2^1, a_2 \cdot 2^2, \ldots$ whose sum constitutes a representation of 1997 in the form under discussion. That is to say, $b(1997)$ is the coefficient of x^{1997} in the product

$$\prod_{i=0}^{10}(1 + x^{2^i} + x^{2^{i+1}}).$$

Calculating this coefficient is duck soup for a modern computer, but a discouragingly tedious undertaking by hand. What a relief it is to discover in the published solution that a few perceptive observations about the function $b(n)$ lead to a virtually effortless resolution of the problem.

The basic points are that $n - a_0$ is an even number,

$$n - a_0 = a_1 \cdot 2 + a_2 \cdot 2^2 + a_3 \cdot 2^3 + \cdots + a_k \cdot 2^k,$$
$$= 2(a_1 + a_2 \cdot 2 + a_3 \cdot 2^2 + \cdots + a_k \cdot 2^{k-1})$$
$$= 2M$$

and that the parenthetic sum here gives a representation of the integer M in the form under discussion.

If n is odd, then a_0 must be 1 and for some nonnegative integer t,

$$n = 2t + 1 = 1 + 2M,$$

and

$$t = M.$$

Thus each representation of $2t + 1$ contains a representation of t. Conversely, to each representation of t there corresponds a representation of $2t + 1$:

> doubling t doesn't have to be achieved by doubling the coefficients a_i, which might push them beyond their prescribed range of 0, 1, 2, but can be accomplished by increasing the exponents in the powers of 2, leaving the coefficients at their acceptable values; appending an initial 1 then gives a representation of $2t + 1$.

Hence the numbers of such representations are the same, and we have the recursion

$$b(2t + 1) = b(t).$$

If n is even, we have

$$n = 2t = a_0 + 2M, \quad \text{where } a_0 = 0 \text{ or } 2.$$

In the case of $a_0 = 0$, we get $t = M$ and have the same situation as above. Hence the number of representations of $2t$ in which $a_0 = 0$ is $b(t)$.

In the case of $a_0 = 2$, we have $2t = 2 + 2M$, and M is a representation of $t - 1$. Conversely, any representation of $t - 1$ can be doubled without altering its coefficients, and when prefaced by a 2 constitutes a representation of $2t$. Thus the number of representations of $2t$ having $a_0 = 2$ is $b(t - 1)$, and altogether we have

$$b(2t) = b(t) + b(t - 1).$$

Now $b(1997)$ is easily found. We have

$$b(1997) = b(998)$$
$$= b(499) + b(498)$$
$$= b(249) + [b(249) + b(248)]$$
$$= 2 \cdot b(249) + b(248)$$
$$= 2 \cdot b(124) + [b(124) + b(123)]$$
$$= 3 \cdot b(124) + b(123)$$
$$= 3[b(62) + b(61)] + b(61)$$
$$= 3 \cdot b(62) + 4 \cdot b(61)$$
$$= 3[b(31) + b(30)] + 4 \cdot b(30)$$
$$= 3 \cdot b(31) + 7 \cdot b(30)$$
$$= 3 \cdot b(15) + 7[b(15) + b(14)]$$
$$= 10 \cdot b(15) + 7 \cdot b(14)$$
$$= 10 \cdot b(7) + 7[b(7) + b(6)]$$
$$= 17 \cdot b(7) + 7 \cdot b(6)$$
$$= 17 \cdot b(3) + 7[b(3) + b(2)]$$
$$= 24 \cdot b(3) + 7 \cdot b(2)$$
$$= 24 \cdot b(1) + 7[b(1) + b(0)]$$
$$= 31 \cdot b(1) + 7 \cdot b(0)$$
$$= 31 \cdot 1 + 7 \cdot 1$$
$$= 38.$$

5. (From "Challenges in Physics and Math" M157, page 24, Nov/Dec, 1995; submitted by L. Kurlyandchik)

$\{a_n\} = \{a_1, a_2, \ldots\}$ is an arbitrary sequence of positive numbers. Let

$$b_n = (a_1 + a_2 + \cdots + a_n)\left(\frac{1}{a_1} + \frac{1}{a_2} + \cdots + \frac{1}{a_n}\right),$$

and

$$c_n = [\sqrt{b_n}], \text{ the integer part of the square root of } b_n.$$

Prove that $c_1, c_2, \ldots$ are all different.

Clearly $b_{n+1} \geq b_n$, and therefore $c_{n+1} \geq c_n$, and if we can show that c_{n+1} and c_n are never the same, it would follow that $c_{n+1} > c_n$, making $\{c_n\}$ a strictly increasing sequence and implying the desired conclusion.

Now, if two real numbers differ by at least 1, they must either be consecutive integers or they straddle an integer. In either case they can't have the same integer part. Hence it suffices to show that

$$\sqrt{b_{n+1}} - \sqrt{b_n} \geq 1,$$

that is, that

$$b_{n+1} \geq b_n + 2\sqrt{b_n} + 1.$$

Since

$$b_{n+1} = \left[(a_1 + a_2 + \cdots + a_n) + a_{n+1}\right]\left[\left(\frac{1}{a_1} + \frac{1}{a_2} + \cdots + \frac{1}{a_n}\right) + \frac{1}{a_{n+1}}\right]$$

$$= b_n + (a_1 + a_2 + \cdots + a_n) \cdot \frac{1}{a_{n+1}}$$

$$+ a_{n+1} \cdot \left(\frac{1}{a_1} + \frac{1}{a_2} + \cdots + \frac{1}{a_n}\right) + 1,$$

we would like to show that

$$\frac{1}{a_{n+1}}(a_1 + a_2 + \cdots + a_n) + a_{n+1}\left(\frac{1}{a_1} + \frac{1}{a_2} + \cdots + \frac{1}{a_n}\right) \geq 2\sqrt{b_n}.$$

But since the numbers are all positive, this follows immediately from the A.M.–G.M. inequality:

$$\left[\frac{1}{a_{n+1}}(a_1 + a_2 + \cdots + a_n)\right] + \left[a_{n+1}\left(\frac{1}{a_1} + \frac{1}{a_2} + \cdots + \frac{1}{a_n}\right)\right]$$

$$\geq 2\sqrt{\left[\frac{1}{a_{n+1}}(a_1 + a_2 + \cdots + a_n)\right] \cdot \left[a_{n+1}\left(\frac{1}{a_1} + \frac{1}{a_2} + \cdots + \frac{1}{a_n}\right)\right]}$$

$$= 2\sqrt{b_n}.$$

6. (From the article "A Pivotal Approach" by Boris Pritsker of New York City, May/June, 1996, page 44)

    Squares $ABPQ$ and $ACRS$ are drawn outwardly on the sides of $\triangle ABC$ (Figure 6). Prove that QS is twice as long as the median AM to BC.

(Another problem concerning this configuration is given in part 3 of section 5.)

Topics Based on Problems from *Quantum*

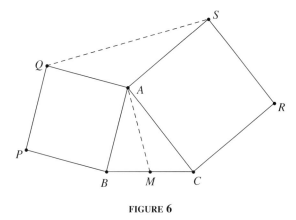

FIGURE 6

In the clockwise rotation through 90° about the center O of square $ABPQ$ (Figure 7), let us look for the image of QS. Clearly Q goes to A. The problem is to determine the image of S.

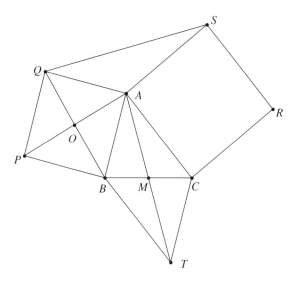

FIGURE 7

To this end, consider the image of AS. Under a rotation, each line is rotated through the angle of rotation. Hence AS is turned to lie in the direction of AC, and since A goes to B and $AS = AC$, the image of AS is the segment BT which is equal and parallel to AC. Hence S goes to T, and since Q goes to A, then

QS goes to AT. Thus $QS = AT$. But AT is a diagonal of parallelogram $ABTC$, and since the diagonals bisect each other, AT is twice the median AM in $\triangle ABC$ and the conclusion follows.

Comment: Since the rotation is through a right angle, we also have the nice result that QS and AM are perpendicular.

(a) While Pritsker's proof is most ingenious and satisfying, I hope you will also enjoy the following simple solution using complex numbers and vectors.

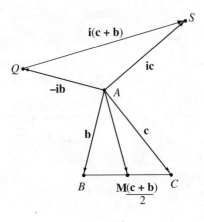

FIGURE 8

Let A be the origin and let vectors **AB** and **AC** be **b** and **c**, respectively (Figure 8). Then, since multiplication by the imaginary unit **i** rotates a vector through a counterclockwise angle of 90° without altering its length, $\mathbf{AS} = \mathbf{ic}$ and $\mathbf{AQ} = \mathbf{i}^3\mathbf{b} = -\mathbf{ib}$. Thus

$$\mathbf{QS} = \mathbf{AS} - \mathbf{AQ} = \mathbf{i}(\mathbf{c} + \mathbf{b}),$$

while

$$\mathbf{AM} = \tfrac{1}{2}(\mathbf{AB} + \mathbf{AC}) = \tfrac{1}{2}(\mathbf{b} + \mathbf{c}).$$

Hence QS is twice as long as AM and they are perpendicular:

the median AM in $\triangle ABC$ lies along the altitude from A in $\triangle AQS$.

(b) Now, as far as the squares are concerned, they don't know whether they've been drawn on sides of △ABC or △AQS. Thus a property of either of these triangles also belongs to the other. Hence, just as the extended median AM gives the altitude from A in △AQS, the extended altitude AD in △ABC bisects QS (Figure 9). I first encountered this latter result independently and I expect there are many proofs of it. I have seen four and I can't resist showing you a most beautiful one that is due to Larry Rice of the University of Toronto Schools.

In Figure 9, let altitude AD meet QS at M and consider the clockwise quarter turn about A. Obviously this takes S to C, and since AM is perpendicular to BC, it is turned to a position AM' which is parallel to BC. Finally, this rotation extends right angle BAQ into a straight angle, carrying AQ to AQ' along the extension of BA, and so QMS is taken to Q'M'C.

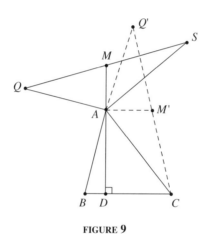

FIGURE 9

Since $BA = AQ = AQ'$, then, in △BQ'C, AM' issues from the midpoint A of BQ', is parallel to BC, and therefore it bisects the third side Q'C. Thus M' is the midpoint of Q'C and it follows that its antecedent M is the midpoint of QS.

7. (Problem M233 from "Challenges in Physics and Math," May/June, 1998; submitted by D. I. Averianov)

Let all the irreducible fractions whose denominators don't exceed 99 be written in increasing order from left to right:

$$\frac{1}{99}, \frac{1}{98}, \ldots, \frac{a}{b}, \frac{5}{8}, \frac{c}{d}, \ldots$$

What are the fractions $\frac{a}{b}$ and $\frac{c}{d}$ on either side of $\frac{5}{8}$?

An ingenious solution is given in the magazine and one doesn't want for more. However, I hope you might enjoy the following simple solution that is available to those who are familiar with the Farey sequences.

The nth Farey sequence, F_n, consists of the row of irreducible fractions from 0 to 1, written from left to right in increasing order, whose denominators are less than or equal to n. For example, F_5 is

$$\frac{0}{1}, \frac{1}{5}, \frac{1}{4}, \frac{1}{3}, \frac{2}{5}, \frac{1}{2}, \frac{3}{5}, \frac{2}{3}, \frac{3}{4}, \frac{4}{5}, \frac{1}{1}.$$

The infinite row of fractions in this problem, then, begins with F_{99}, and the three fractions in question, $\frac{a}{b}, \frac{5}{8}, \frac{c}{d}$ are well contained in F_{99}.

(a) Now, it is known, for consecutive fractions $\frac{a}{b}$ and $\frac{c}{d}$ in a Farey sequence, that

$$bc - ad = 1.$$

(This is proved in my *Ingenuity in Mathematics*, The Anneli Lax New Mathematical Library Series, MAA, vol. 23, 1970.)

Hence, for $\frac{a}{b}$ immediately preceding $\frac{5}{8}$ in a Farey sequence, we have

$$5b - 8a = 1,$$
$$8a + 1 = 5b \equiv 0 \pmod{5},$$

implying

$$a \equiv 3 \pmod{5}.$$

Hence, for some positive integer n, $a = 5n + 3$ and, substituting in $5b - 8a = 1$, we get

$$5b - 40n - 24 = 1, \quad b = 8n + 5,$$

and

$$\frac{a}{b} = \frac{5n+3}{8n+5}.$$

That is to say, the neighbor which is immediately to the left of $\frac{5}{8}$ in a Farey sequence is a fraction $\frac{5n+3}{8n+5}$. Since every Farey sequence F_r is contained in every later sequence F_{r+k}, the neighbor immediately to the left of $\frac{5}{8}$ will remain its neighbor until some greater fraction intervenes in a later sequence; these neighbors are nondecreasing as one goes from sequence to sequence.

Now, it is easy to see that $\frac{5n+3}{8n+5}$ increases with n:

$$\frac{5(n+1)+3}{8(n+1)+5} - \frac{5n+3}{8n+5} = \frac{5n+8}{8n+13} - \frac{5n+3}{8n+5}$$

$$= \frac{1}{(8n+13)(8n+5)} > 0.$$

Next, let us show that $\frac{5n+3}{8n+5}$ is in its lowest terms:

every common divisor d of $8n+5$ and $5n+3$ also divides $5(8n+5) - 8(5n+3) = 1$, implying $d = 1$.

Thus $8n+5$ is the actual value of the denominator in the sequence, and since it must remain ≤ 99, its greatest value is 93, given by $n = 11$. That is to say, up to F_{99}, the neighbors of $\frac{5}{8}$ run through the 11 increasing values of $\frac{5n+3}{8n+5}$ for $n = 1, 2, \ldots, 11$. The current neighbor in F_{99} is therefore

$$\frac{5 \cdot 11 + 3}{8 \cdot 11 + 5} = \frac{58}{93}.$$

($\frac{58}{93}$ started out in this role in F_{93}, and will remain so until F_{101}, when it will yield to $\frac{5 \cdot 12 + 3}{8 \cdot 12 + 5} = \frac{63}{101}$.)

(b) Similarly for the neighbor $\frac{c}{d}$ immediately to the right of $\frac{5}{8}$.
We have $8c - 5d = 1$,

$$5d = 8c - 1, \quad 8c \equiv 1 \pmod{5}, \quad c = 5n+2;$$

then

$$5d = 40n + 15, \quad d = 8n+3,$$

and

$$\frac{c}{d} = \frac{5n+2}{8n+3} \quad \text{(in its lowest terms)}.$$

We observe that this time, as one goes from sequence to sequence, this neighbor either remains the same or gets smaller in order to get closer to $\frac{5}{8}$; and it is easy to check that $\frac{5n+2}{8n+3}$ decreases as n increases. Thus it is again the greatest value of n that gives the current neighbor in F_{99}. Restricting

$8n + 3 \leq 99$ allows $n = 12$ and the denominator 99. Hence the required neighbor is

$$\frac{5 \cdot 12 + 2}{8 \cdot 12 + 3} = \frac{62}{99}.$$

8. Now for an innocent question in arithmetic (given by I. F. Sharygin in the article "So, What's Wrong" in July/August 1998, page 34)

A farmer harvested ten tons of watermelons and sent them by river to the nearest town. It is well known that a watermelon, as reflected in its name, is made almost entirely of water. When the barge left, the content of the watermelons was 99% water by weight. On the way to town, the watermelons dried out somewhat and their water content dropped to 98%.

What was the weight of the watermelons when they reached the town?

You would swear that the watermelons would only have to dry out a little in order to drop the water content the bit from 99% to 98%. Would you believe that it takes a loss of five tons of water to do this?

The way to look at things is to consider the non-water part of the melons—the pulp. At the beginning, the pulp constituted 1% of the total ten tons—that would be 200 pounds. At the end, this 200 pounds of pulp constituted 2% of what was left, in which case 100% of the remaining total is 50 times as much, that is, $50 \cdot 200 = 10,000$ pounds. That's an amazing five tons!

9. Here is another Sharygin offering. (Problem M252, Jan./Feb. 1999, page 23)

Two nonintersecting circles are tangent to both arms of acute angle $\angle XOY = \alpha$ (Figure 10). Construct an isosceles triangle ABC with ver-

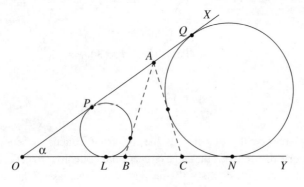

FIGURE 10

tex A on OX and base BC on OY such that each of its equal sides is tangent to one of the circles.

Professor Sharygin's solution gives a wonderfully simple construction for $\triangle ABC$ based on the pleasing result that the altitude to its base is just the sum of the radii. What follows can't compete with that, but if all you want to do is to construct the triangle, it might be of interest.

Suppose the triangle has been constructed and that AD is the altitude to the base. Since $\triangle ABC$ is isosceles, D is the midpoint of BC. The two tangents to a circle from a point outside are the same length, and the equal pairs are marked a, b, c, d, e, in Figure 11, where each half of BC is marked t.

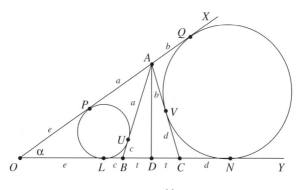

FIGURE 11

Let $\triangle ABC$ be cut at U and V and the parts of its equal sides folded out along OX and OY. In this way the perimeter of $\triangle ABC$ is copied onto the segments PQ and LN. Since $PQ = LN$, by symmetry, then the length of each of these segments is the semiperimeter s of $\triangle ABC$. Now, AD also bisects the perimeter and we have $AC + CD = s$. Therefore

$$AC + CD = LN,$$

and we have

$$b + d + t = c + 2t + d,$$

giving

$$b = c + t = LD.$$

Thus

$$AO + OD = a + e + e + b = (b + a + e) + e = QO + OL.$$

The points of contact Q and L are easily determined and provide a simple construction of a segment of length $(QO + OL)$. Since $AO + OD = QO + OL$, if a segment of length $(OQ + OL)$ is laid along XO, bent at O to run along OY, and slid around until the segment joining the endpoints is perpendicular to OY, the endpoints will wind up at A and D.

From $\triangle AOD$ we have

$$\frac{OD}{AO} = \cos \alpha.$$

Thus, in order to find out where to bend the segment around O, one need only divide it in the ratio $m : n$, where $m/n = \cos \alpha$ (Figure 12). Now, $\cos \alpha$ is readily found as the ratio of two segments m and n by erecting a perpendicular FG to OX from any point F on OY (Figure 13). Thus vertex A may be determined and the rest of the triangle easily completed.

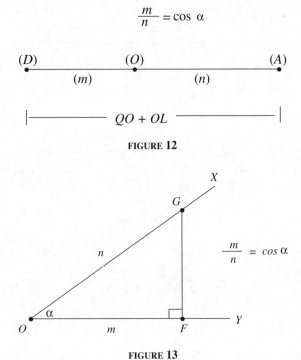

FIGURE 12

FIGURE 13

10. (Problem M269 from the regular column "Challenges," July/August, 1999, page 11)

If x and y are positive integers and the sum of the fractions

$$\frac{x^2-1}{y+1} + \frac{y^2-1}{x+1}$$

is an integer, prove that each of the fractions must also be an integer. We follow the published solution.

Since x and y are positive integers, at least the fractions are rational. Now, the key is to observe that the *product* of the fractions is an integer:

$$\frac{x^2-1}{y+1} \cdot \frac{y^2-1}{x+1} = (x-1)(y-1).$$

Thus, if we call the fractions u and v, we have both $u+v$ and uv are integers. Letting $u+v = -m$, and $uv = n$, then u and v are the roots of the equation

$$z^2 + mz + n = 0: \quad u, v = \frac{-m \pm \sqrt{m^2 - 4n}}{2}.$$

Hence $\sqrt{m^2 - 4n}$ must reduce to some integer k in order for these roots to be rational, and

$$u, v = \frac{-m \pm k}{2}.$$

Now, if k has the same parity as m, then the numerator will be an even number regardless of which sign is taken, and hence divisible by 2, giving integer roots, as desired. It is easy to see that this is the case:

clearly $k = \sqrt{m^2 - 4n}$ has the same parity as $m^2 - 4n$, and since $4n$ is even, that's the parity of m^2 and m.

11. (Problem M271 from the regular column "Challenges," September/October, 1999, page 9)

Solve the system of equations

$$x + [y] + \{z\} = 3.9 \tag{1}$$
$$y + [z] + \{x\} = 3.5 \tag{2}$$
$$z + [x] + \{y\} = 2, \tag{3}$$

where $[a]$ and $\{a\}$ denote, respectively, the integer and fractional parts of the real number a:

$$a = [a] + \{a\}.$$

There are really nine numbers to be concerned with here: x, $[x]$, $\{x\}$, y, $[y]$, $\{y\}$, z, $[z]$, $\{z\}$. Thus we might have wished for nine equations; but that isn't the happiest of thoughts either. Of course, we can always draw on the three equations like $a = [a] + \{a\}$, and we also know that $[a]$ is an integer and $\{a\}$ is less than 1.

To begin, however, we need to observe that, while a and $[a]$ might be negative, $\{a\}$ is never less than zero. Consider, for example, $a = -3.9$. Then the greatest integer not exceeding a is $[a] = -4$, and

$$\{a\} = a - [a] = -3.9 - (-4) = 0.1.$$

We will need to know that $0 \leq \{a\} < 1$.

Also, we don't need to isolate each of these unkowns in an equation of its own. Because of their characteristics, an equation like

$$[a] + \{b\} = 3.7$$

implies both

$$[a] = 3 \quad \text{and} \quad \{b\} = 0.7.$$

Armed with these notions, the system gives up without much of a struggle.

Adding equations (2) and (3) and subtracting (1) gives

$$\bigl(y + [z] + \{x\}\bigr) + \bigl(z + [x] + \{y\}\bigr) - \bigl(x + [y] + \{z\}\bigr) = 1.6,$$

which reduces to

$$2\{y\} + 2[z] = 1.6,$$

and

$$\{y\} + [z] = 0.8.$$

Hence

$$\{y\} = 0.8 \quad \text{and} \quad [z] = 0.$$

Similarly, $(1) + (2) - (3)$ gives

$$\bigl(x + [y] + \{z\}\bigr) + \bigl(y + [z] + \{x\}\bigr) - \bigl(z + [x] + \{y\}\bigr) = 5.4,$$
$$2\{x\} + 2[y] = 5.4,$$
$$\{x\} + [y] = 2.7,$$

and

$$\{x\} = 0.7 \quad \text{and} \quad [y] = 2.$$

Finally, (1) + (3) − (2) gives

$$(x + [y] + \{z\}) + (z + [x] + \{y\}) - (z + [x] + \{y\}) = 1.2,$$
$$2[x] + 2\{z\} = 2.4,$$
$$[x] + \{z\} = 1.2,$$

and

$$[x] = 1, \quad \{z\} = 0.2.$$

Hence

$$x = 1.7, \quad y = 2.8, \quad \text{and} \quad z = \{z\} = 0.2.$$

12. There always seems to be a distance-rate-time problem with a new wrinkle. Here is an easy revision of Problem M281 from the regular column "Challenges," January/February, 2000, page 21.

A professor and her assistant like to go for a stroll in the evening and they always walk back and forth along the path between their houses, each beginning at home. They walk at uniform rates, but the assistant is faster.

One evening they left home at the same time and passed each other at a point 55 meters from the professor's house. After turning around at the professor's house, the assistant caught up with the professor before she reached her house at a point 85 meters from the house.

Now, there is a news-stand 25 meters from the assistant's house and an ice cream stand not far from the professor's house. It is known that, at the beginning of their walks on the evening in question, they passed the stands near their houses at the same time.

How far apart are the news-stand and the ice cream stand?

The solution is straightforward. Let P, I, N, and A denote, respectively, the professor's house, the ice cream stand, the news-stand, and the assistant's house (Figure 14). Suppose the assistant walks r times as quickly as the professor, and hence r times as far in a given time. Thus $AN = r \cdot PI$, and since $AN = 25$, then $PI = \frac{25}{r}$, making the required distance

$$IN = PA - 25 - \frac{25}{r}.$$

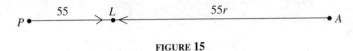

FIGURE 14

Similarly, when the professor walks the 55 meters from P to L in Figure 15, the assistant has walked $55r$ meters from A to L, and we have $PA = 55 + 55r$. Hence

$$IN = 30 + 55r - \frac{25}{r},$$

and it remains to determine r.

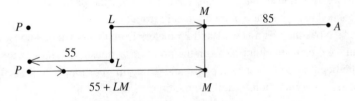

FIGURE 15

Let M be the point where the assistant catches up to the professor (Figure 16). Then, as the professor goes from L to M, the assistant walks from L to P and back to M, and we have

$$r \cdot LM = LP + PM = 55 + (55 + LM) = 110 + LM,$$

and

$$(r - 1)LM = 110.$$

FIGURE 16

But

$$LM = LA - MA = 55r - 85,$$

and we have

$$(r - 1)(55r - 85) = 110,$$
$$(r - 1)(11r - 17) = 22,$$

giving
$$11r^2 - 28r - 5 = 0.$$
Hence
$$r = \frac{28 \pm \sqrt{784 + 220}}{22},$$
where the + sign must be taken to yield a positive result. Thus
$$r = \frac{14 + \sqrt{251}}{11},$$
and
$$\frac{25}{r} = \frac{25 \cdot 11}{14 + \sqrt{251}} \cdot \frac{\sqrt{251} - 14}{\sqrt{251} - 14}$$
$$= \frac{25 \cdot 11(\sqrt{251} - 14)}{55} = 5(\sqrt{251} - 14).$$
Hence
$$IN = 30 + 55\left(\frac{14 + \sqrt{251}}{11}\right) - 5(\sqrt{251} - 14)$$
$$= 30 + 5\left(14 + \sqrt{251}\right) - 5\left(\sqrt{251} - 14\right)$$
$$= 170.$$

13. (Problem B293 from the regular column "Just for the fun of it!," May/June, 2000, page 3)

A man fills up two tanks with water from two hoses. The one hose delivers water at the rate of 2.9 litres per minute and the other at the rate of 8.7 litres per minute. He puts a hose in each tank and starts filling them up at the same time. When the smaller tank was half full he switched hoses and found that the two tanks filled up completely at the same time. If the small tank holds 12.6 litres, how big is the larger tank?

Observe that the one hose is three times as fast as the other and that the smaller tank gets half its water, i.e., 6.3 litres, from each hose.
While the small tank is getting 6.3 litres from the fast hose, the big one gets only one-third as much from the slow hose, i.e., 2.1 litres, and while the small tank is getting 6.3 litres from the slow hose, the big one gets three times

as much from the fast hose, i.e., 18.9 litres. Thus the capacity of the big tank is $2.1 + 18.9 = 21$ litres.

14. Finally let us consider the intriguing sequence of problem M185, proposed by A. Shapovalov, in the regular column "Challenges in Physics and Math," Sept/Oct, 1996.

Does there exist a sequence of positive integers containing each positive integer exactly once such that the sum of the first k terms is divisible by k for each $k = 1, 2, 3, \ldots$?

First let me tell you about my high hopes that ended in failure. It seemed promising to try to construct such a sequence $\{a_1, a_2, a_3, \ldots\}$ by starting with $a_1 = 1$ and thereafter putting in the *smallest* positive integer a_k that is not already in the sequence and which makes the partial sum $S_k (= a_1 + a_2 + \cdots + a_k)$ divisible by k. This led to a sequence which begins

$$1, 3, 2, 6, 8, 4, 11, 5, 14, 16, 7, 19, 21, 9, 24,$$
$$10, 27, 29, 12, 32, 13, 35, 37, 15, 40.$$

Since this didn't give away any secrets, I pinned my hopes on the chance that the partial sums and the quotients $n_k = S_k/k$ might reveal something. So I proceeded to construct the following table.

a_k	1	3	2	6	8	4	11	5	14	16	7	19	21
S_k	1	4	6	12	20	24	35	40	54	70	77	96	117
k	1	2	3	4	5	6	7	8	9	10	11	12	13
n_k	1	2	2	3	4	4	5	5	6	7	7	8	9

a_k	9	24	10	27	29	12	32	13	35	37	15	40	—
S_k	126	150	160	187	216	228	260	273	308	345	360	400	—
k	14	15	16	17	18	19	20	21	22	23	24	25	—
n_k	9	10	10	11	12	12	13	13	14	15	15	16	—

While a_k fluctuates widely, n_k seems to increase through the positive integers, with no integer occurring more than twice. Also, the symmetry

$$a_k = n, \quad a_n = k,$$

holds as far as the table goes; for example, $a_4 = 6$ and $a_6 = 4$. Since k runs through all positive integers, this symmetry implies that every integer $k(= a_n)$ occurs in the sequence. Furthermore, it also guarantees uniqueness, for if $a_n = a_m = k, m > n$, then the contradiction $a_k = n$ and $a_k = m$ would follow. Now, this is all well and good, but unfortunately I wasn't able to establish the symmetry.

I did notice that the table gave support to the conjecture

$$a_{k+1} = \begin{cases} n_k & \text{if } n_k \text{ is not already in the sequence,} \\ n_k + (k+1) & \text{if } n_k \text{ is already in the sequence.} \end{cases}$$

But this is as far as I got.

Granted that my choice to define the sequence in terms of the minimum positive integer that is needed to make n_k an integer might be a rather natural approach, it has the disadvantage of not being easy to apply in establishing other properties of the sequence. In the published solution, which is splendid work by A. Shapovalov and O. Lyashko, the above conjecture, along with $a_1 = 1$, is taken as the definition of a sequence that they propose to show has the desired property. This is a very strategic move because it provides them with a valuable tool. Since integers are not mentioned in the definition, this approach introduces the extra requirement of showing n_k to be a positive integer. But this really doesn't involve extra work since it is equivalent to the central divisibility property that S_k is divisible by k (recall the definition $n_k = S_k/k$).

(a) n_k is a positive integer.

This follows easily by induction. Clearly $n_1 = \frac{S_1}{1} = \frac{1}{1} = 1$ is a positive integer. Suppose, for some $k \geq 1$, n_k is a positive integer. Then, in the event that $a_{k+1} = n_k$, we have

$$n_{k+1} = \frac{S_{k+1}}{k+1} = \frac{S_k + a_{k+1}}{k+1} = \frac{k \cdot n_k + n_k}{k+1} = n_k;$$

and if $a_{k+1} = n_k + (k+1)$, then

$$n_{k+1} = \frac{S_{k+1}}{k+1} = \frac{k \cdot n_k + n_k + (k+1)}{k+1} = n_k + 1,$$

again a positive integer in either case.

This argument also yields the following three corollaries:

(i) the sequence $\{n_k\}$ is nondecreasing, and in particular, that
(ii) $n_{k+1} = n_k$ iff $a_{k+1} = n_k$,
$n_{k+1} = n_k + 1$ iff $a_{k+1} = n_k + (k+1)$.
(iii) The equality $n_{k+1} = n_k$, in implying the first option for the value of a_{k+1}, namely $a_{k+1} = n_k$, reveals that n_k does not occur earlier in the sequence than a_{k+1}:

$$n_{k+1} = n_k \text{ implies that } n_k \text{ does not occur ahead of } a_{k+1}.$$

(b) Continuing to consider the behaviour of the sequence $\{n_k\}$ (my cherished symmetry never does make an appearance in the solution), it is not difficult to establish that n_k takes on all positive integral values. This follows from the nondecreasing character of $\{n_k\}$ and

(iv) $n_{k+2} > n_k$ ($\{n_k\}$ increases at least with every second term).

Proof of (iv): Since $\{n_k\}$ is nondecreasing, $n_{k+2} \geq n_{k+1} \geq n_k$. Thus, if n_{k+2} were to have the same value as n_k, it would follow that

$$n_{k+2} = n_{k+1} = n_k = \text{some positive integer } m.$$

Now, by (iii) above, $n_{k+2} = n_{k+1}$ implies that $n_{k+1}(= m)$ does not occur ahead of a_{k+2}. However, by (ii), $n_{k+1} = n_k$ implies that $a_{k+1} = n_k = m$, and we have a contradiction already.

Thus n_k makes its way through all the positive integers.

Now we can argue easily to establish the remaining two properties of the sequence.

(c) At each stage, either n_k is already in the sequence or it goes in as a_{k+1}. Thus all values of n_k, that is, all the positive integers, occur in the sequence.

(d) Finally, it remains only to show that no integer occurs more than once.

Suppose that some term a_m occurs again later as a_k. Consider the options for the value of a_k, namely $a_k = n_{k-1}$ or $a_k = n_{k-1} + k$. If a_k were to come into the sequence as n_{k-1}, then $a_m (= a_k)$ would constitute an earlier occurrence of n_{k-1}, in which case the rule of formation requires a_k to be assigned the value $n_{k-1} + k$. Hence a_k must have entered the sequence as the integer $n_{k-1} + k$.

However, $a_k = n_{k-1} + k$ is too big to be the same as a_m, which is either n_{m-1} or $n_{m-1} + m$. Since $\{n_k\}$ is nondecreasing, $n_{k-1} \geq n_{m-1}$, and since k is strictly greater than m, we have

$$a_k = n_{k-1} + k > n_{m-1} + m \geq a_m,$$

a final contradiction.

SECTION 9
Two Distinguished Integers*

1. Let n be a positive integer and S the set of positive integers between 1 and n that are relatively prime to n. The table below shows the sets S for a few values of n. Observe that, for $n = 12$, S contains only prime numbers, and that this is also the case for $n = 18$ and $n = 30$. To our surprise, however, this never happens again:

30 is the greatest positive integer having a set S that is comprised exclusively of prime numbers.

n	S
10	3, 7, 9
12	**5, 7, 11**
15	2, 4, 7, 8, 11, 13, 14
18	**5, 7, 11, 13, 17**
26	3, 5, 7, 9, 11, 15, 17, 19, 21, 23, 25
30	**7, 11, 13, 17, 19, 23, 29**

Let p_m denote the mth prime number. It is not difficult to prove that, beyond the second prime, no prime is as big as the product of all the preceding primes:

$$p_{m+1} < p_1 p_2 \cdots p_m.$$

Using the ingenious approach that Euclid employed to show that there exists an infinity of primes, consider the number

$$A = p_1 p_2 \cdots p_m - 1, \quad m \geq 2.$$

Clearly A is bigger than 1 and therefore must have some prime divisor p. However, each of the primes $p_1, p_2, \ldots, p_m$ leaves a remainder of -1 when

---

*This is a reworked version of material from my "Mathematical Gems" column in the *Two-Year College Mathematics Journal*, Vol. 10, June, 1979.

divided into A. Thus every prime divisor p of A must be at least as great as p_{m+1}. But no divisor of A can exceed A, and we have the desired

$$p_{m+1} \leq p \leq A < p_1 p_2 \cdots p_m.$$

A pretty proof of the above property of the number 30 can be based on the similar inequality

$$p_{m+1}^2 < p_1 p_2 \cdots p_m,$$

which is valid for $m \geq 4$. This represents a marked improvement on the inequality just established, but its proof, although entirely elementary, would sidetrack us too long. The proof is given in full detail on pages 187–192 of *The Enjoyment of Mathematics* by Rademacher and Toeplitz, a book which it is impossible to praise too highly. This first part of our story is based on material in this wonderful book.

Let us assume, then, that $p_{m+1}^2 < p_1 p_2 \cdots p_m$ for $m \geq 4$ (clearly $49 = p_4^2$ is not less than $30 = 2 \cdot 3 \cdot 5$). Since the square of a prime is a composite number, an exclusively prime set S cannot contain any of the squares 4, 9, 25, 49, 121, 169, 289, ... In order to avoid having $p_m^2 (< n)$ in its set S, an integer n must be divisible by p_m so that n and p_m^2 will fail to be relatively prime. Thus $n > 49$ must be divisible by 2, 3, 5, and 7, in order to keep out 4, 9, 25, and 49. Hence

$$n \geq 2 \cdot 3 \cdot 5 \cdot 7 = 210.$$

Unfortunately, this takes n beyond $11^2 = 121$, which brings with it the need to be divisible by 11. Altogether, then, n must be at least as great as

$$2 \cdot 3 \cdot 5 \cdot 7 \cdot 11 = 2310.$$

And so it goes: exceeding 169, n must be divisible by 13, making it at least $13 \cdot 2310 = 30030$, shooting it past 289, ... To summarize,

in escaping the square $p_m^2 (< n)$ and all the smaller squares p_i^2, it is necessary that $n \geq p_1 p_2 \cdots p_m > p_{m+1}^2$, which, in turn, requires n to be divisible by p_{m+1}, and leads to $n > p_{m+2}^2$, etc., etc. Clearly no finite n can continue to meet these requirements indefinitely, and must admit some composite p_m^2 into its set S somewhere down the line.

It remains to consider the integers from 31 to 49. Since each exceeds the square 25, it must be divisible by $2 \cdot 3 \cdot 5 = 30$ in order to exclude 4, 9, and 25 from S. Since none of these qualify, the proof is complete.

2. Since $\sqrt{12}$ lies between 3 and 4, the set of positive integers less than $\sqrt{12}$ consists of just 1, 2, and 3, each of which we observe is a divisor of 12. Again, the positive integers less than $\sqrt{24}$ are 1, 2, 3, and 4, and each is a divisor of 24. Beyond 24, however, this never happens again:

24 is the greatest positive integer that is divisible by each of the positive integers less than its square root.

I expect you must be thinking that we have gone on to a new topic. If so, you might be pleasantly surprised to learn that this result is an easy corollary to the property of the number 30 which we just proved.

Since 5 does not divide 26, 27, 28, or 29, and 4 does not divide 25 or 30, it is clear that any integer n, greater than 24, that possesses the property in question must exceed 30. However, for $n > 30$, its set S of smaller relatively prime integers must contain a composite integer $m = ab < n$.

Now, not both the factors a and b can exceed $\sqrt{m}$, lest their product ab exceed m. For definiteness, suppose $b \leq \sqrt{m}$. Then, since $m < n$, we have

$$b = \sqrt{m} < \sqrt{n}, \quad \text{that is,} \quad b < \sqrt{n}.$$

But, since m belongs to S, m and n are relatively prime. The divisor b of m, then, is also relatively prime to n, and hence b must fail to divide n.

SECTION 10
A Property of the Binomial Coefficients*

Given a prime p, there are certain positive integers n such that none of the binomial coefficients of rank n is divisible by p, that is, none of

$$\binom{n}{0}, \binom{n}{1}, \binom{n}{2}, \ldots, \binom{n}{n}$$

is divisible by p. For example, if $p = 7$, such a number is $n = 20$, for none of

$$\binom{20}{0}, \binom{20}{1}, \binom{20}{2}, \ldots, \binom{20}{20}$$

is divisible by 7. We shall see that there are, in fact, infinitely many such n for each prime p. The problem of determining the values of n for a given prime was posed in 1976 by M. R. Raikar and M. R. Modak of S. P. College, Poona, India. Their straightforward solution follows.

Suppose n is a positive integer such that no $\binom{n}{r}$ is divisible by p. Now, every positive integer is either a power of p or a number between two consecutive powers of p. Suppose

$$p^k \le n < p^{k+1}.$$

Then, in evaluating the coefficients $\binom{n}{r}$, as r runs through the numbers $0, 1, 2, \ldots, n$, we eventually reach $r = p^k$. Consider $\binom{n}{r}$ when $r = p^k$.

It is easy to derive the general relation

$$r\binom{n}{r} = (n - r + 1)\binom{n}{r-1},$$

implying that r divides $(n - r + 1)\binom{n}{r-1}$. For $r = p^k$, then, we have

$$p^k \mid (n - r + 1)\binom{n}{r-1}.$$

---

*This is a reworked version of material from my "Mathematical Gems" column in the *Two-Year College Mathematics Journal*, Vol. 11, March, 1980.

Since, by assumption, no $\binom{n}{r}$ is divisible by p, this gives
$$p^k \mid (n - r + 1),$$
and since $r = p^k$, that
$$p^k \mid n + 1.$$
Thus, for some positive integer t,
$$n + 1 = tp^k.$$
Now, because $n \geq p^k$, we have
$$tp^k = n + 1 > p^k,$$
implying
$$t > 1.$$
On the other hand,
$$tp^k = n + 1 \leq p^{k+1},$$
and
$$t \leq p.$$
Thus
$$t \in \{2, 3, \ldots, p\},$$
and, for n in the above interval between p^k and p^{k+1}, the only candidates for the value of $n + 1$ are
$$2p^k, 3p^k, \ldots, p \cdot p^k.$$

Over the whole range of positive integers, then, the entire list of candidates for $n + 1$ is
$$\left[(2, 3, \ldots, p), (2p, 3p, \ldots, p \cdot p), (2p^2, 3p^2, \ldots, p \cdot p^2), \ldots\right].$$
We may regroup these numbers slightly by making the last one in each group the first one in the next group:
$$\Big[(2, 3, \ldots, p - 1), (p, 2p, 3p, \ldots, (p - 1)p),$$
$$(p^2, 2p^2, 3p^2, \ldots, (p - 1)p^2), \ldots\Big].$$

A Property of the Binomial Coefficients

Thus, for the number n itself, we have the candidates

$$n = tp^m - 1, \quad \text{where} \quad m \geq 0 \quad \text{and} \quad t = 1, 2, \ldots, p-1.$$

(For $m = 0$ and $t = 1$, this summary allows $n = 0$, giving the candidate $n + 1 = 1$, which is not in our list. However, since the only value of $\binom{0}{r}$ is $\binom{0}{0} = 1$, which is not divisible by p, it is a harmless addition.)

At this point, we have no idea how many of these candidates $tp^m - 1$ actually provide a value of n such that no $\binom{n}{r}$ is divisible by p. We only know that any such n must occur among them. What a nice surprise it is, therefore, to learn that every one of these numbers is an acceptable value of n.

Suppose, then, that n is given by

$$n = tp^m - 1, \quad \text{where} \quad m \geq 0 \quad \text{and} \quad t \in \{1, 2, \ldots, p-1\}.$$

We shall show by induction on r that none of

$$\binom{n}{0}, \binom{n}{1}, \binom{n}{2}, \ldots, \binom{n}{n}$$

is divisible by p.

Since $\binom{n}{0} = 1$, p does not divide $\binom{n}{r}$ for $r = 0$. In the case of $\binom{n}{r}$, where $r \geq 1$, suppose that p does not divide the previous binomial coefficient $\binom{n}{r-1}$. We noted earlier that

$$r\binom{n}{r} = (n - r + 1)\binom{n}{r-1}.$$

Since p does not divide $\binom{n}{r-1}$, the only factor on the right side which might be divisible by p is $(n - r + 1)$. Let the greatest power of p that divides $(n - r + 1)$ be p^q. We shall show that, on the left side, $r\binom{n}{r}$, all q of these factors p are contributed by the factor r, leaving none for the factor $\binom{n}{r}$, and implying that p does not divide $\binom{n}{r}$.

Since $n - r + 1 \leq n + 1 = tp^m < p^{m+1}$ (recall $t < p$), the factor p^q of $(n - r + 1)$ cannot exceed p^m. In this case p^q must divide p^m, implying that it also divides tp^m. That is to say,

$$p^q \text{ divides } n + 1.$$

Hence $n - r + 1$ and $n + 1$ are both multiples of p^q, and therefore p^q divides their difference

$$n + 1 - (n - r + 1) = r,$$

completing the proof.

As a final observation, note that for $p = 2$, the only value of t is 1, and the acceptable $n = 2^m - 1$. Thus the values of $\binom{n}{r}$ are all odd if and only if $n = 2^m - 1$.

Now, if $\binom{n}{r}$ is odd, the famous Pascal recursion,

$$\binom{n-1}{r-1} + \binom{n-1}{r} = \binom{n}{r},$$

implies that

$$\binom{n-1}{r-1} \quad \text{and} \quad \binom{n-1}{r}$$

must have opposite parity. Hence the binomial coefficients $\binom{2^m-2}{r}$ must alternate odd, even, odd, even, ...

SECTION 11
Nine Miscellaneous Problems

1. (From the 1990 Israeli-Hungarian Competition)

   Prove that x^2+y+2 and y^2+4x are never both perfect squares for the same pair of positive integers (x, y). (For example, for $(1, 6)$, $x^2 + y + 2 = 3^2$ while $y^2 + 4x = 40$, and for $(2, 1)$, $y^2 + 4x = 3^2$ while $x^2 + y + 2 = 7$.)

   As usual, when required to prove something cannot happen, let us proceed indirectly. Suppose

   $$x^2 + y + 2 = s^2 \quad \text{and} \quad y^2 + 4x = t^2.$$

Since x and y are positive, s must be greater than x, and t is greater than y. Hence, for some positive integers a and b, we have

$$x^2 + y + 2 = (x + a)^2 = x^2 + 2ax + a^2$$

and

$$y^2 + 4x = (y + b)^2 = y^2 + 2by + b^2,$$

in which we reap the enormous advantage of having the terms in x^2 and y^2 cancel, yielding the simple system

$$2ax - y = 2 - a^2, \tag{1}$$
$$4x - 2by = b^2. \tag{2}$$

Solving, $2b$ times (1) gives

$$4abx - 2by = b(4 - 2a^2) \tag{3}$$

and $(3) - (2)$ is

$$4(ab - 1)x = b(4 - 2a^2 - b),$$

87

from which
$$x = \frac{b(4 - 2a^2 - b)}{4(ab - 1)}.$$

But this expression is never a positive integer for any positive integers a and b:

$a = b = 1$ must be avoided in order to keep the denominator from vanishing, in which case one of a and b is at least 2, and the other still at least 1. This makes the denominator positive, which in turn requires the numerator to be positive. With b positive, the factor $4 - 2a^2 - b$ must also be positive; but this never happens when one of a and b is 2 or more and the other at least 1.

The conclusion follows.

Exercise Find all pairs of positive integers (x, y) for which
$$x^2 + 3y \quad \text{and} \quad y^2 + 3x$$
are both perfect squares.

2. (From a Bulgarian Competition for Year 11)

Find the minimum value of the function
$$f(x) = \sqrt{x^2 - x + 1} + \sqrt{x^2 - \sqrt{3}x + 1}.$$

One who has a geometric turn of mind might observe that, by completing the squares, the function may be written as
$$\sqrt{\left(x - \frac{1}{2}\right)^2 + \left(\frac{\sqrt{3}}{2}\right)^2} + \sqrt{\left(\frac{\sqrt{3}}{2} - x\right)^2 + \left(\frac{1}{2}\right)^2},$$
which is the sum of the lengths of the hypotenuses of abutting right triangles CAP and PBD in Figure 1.

Thus, for all x, $f(x)$ is given by the length of a polygonal path CPD, between fixed points C and D. Note that the length of AB is constant as x and P vary:
$$AB = \left(x - \frac{1}{2}\right) + \left(\frac{\sqrt{3}}{2} - x\right) = \frac{\sqrt{3} - 1}{2}.$$

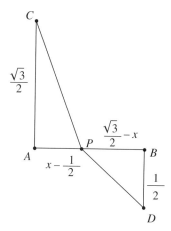

FIGURE 1

Clearly, $CP + PD$ is a minimum when CPD is straight. Thus the required minimum is the length of the hypotenuse of right triangle CED, where CE is perpendicular to the extension of BD (Figure 2).

Observing that

$$CE = AB = \frac{\sqrt{3}-1}{2} \quad \text{and} \quad BE = AC = \frac{\sqrt{3}}{2},$$

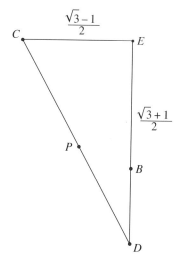

FIGURE 2

we have
$$DE = \frac{\sqrt{3}}{2} + \frac{1}{2} = \frac{\sqrt{3}+1}{2},$$
and obtain
$$\min f(x) = \sqrt{\left(\frac{\sqrt{3}-1}{2}\right)^2 + \left(\frac{\sqrt{3}+1}{2}\right)^2} = \sqrt{2}.$$

3. (From the 1998 AHSME [American High School Mathematics Examination])

Three cards, each with one of the positive integers x, y, z written on it, are lying face-down on a table. The numbers
   (i) are all different,
   (ii) add up to 13,
   (iii) are in increasing order from left to right: $x < y < z$.

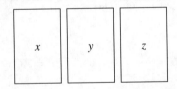

Three girls, A, B, and C, who know conditions (i), (ii), (iii), act as follows.

First A takes a private look at x, and after thinking about it for awhile announces that she is unable to figure out what the three numbers are.

Next B takes a private look at z, and after some thought also announces that she doesn't know what the numbers are.

Finally, C looks privately at y, and also announces that she can't figure out what the numbers are.

Now, these are three smart girls who aren't trying to fool anybody. The question is "What's the number on the middle card?"

First of all, if $x \geq 4$, then the sum $S = x + y + z \geq 4 + 5 + 6 = 15$, exceeding its value of 13. Hence x must be either 1, 2, or 3.

Furthermore, if $x = 3$, then A could deduce that the numbers would have to be $(3, 4, 6)$:

if y were as big as 5, then $S \geq 3 + 5 + 6 > 13$;
thus y, being bigger than x, could only be 4, in which case $z = 6$.

Hence, in order to stump A, x must be either 1 or 2.

Now, B and C have also figured this out just as we did, and so they have more information to work on when it's their turn to look at a card.

Clearly z cannot be 11 or more without violating $S = 13$.

If $z = 10$, then the numbers, being different, would have to be $(1, 2, 10)$, and B would be aware of this. Hence z can't be more than 9.

Also, $z = 9$ sets the numbers at $(1, 3, 9)$:

if x were to be 2, then $y \geq 3$, making $S \geq 2 + 3 + 9 > 13$; thus $x = 1$, and it follows that $y = 3$.

Since B can't determine the numbers, then $z \leq 8$.

Again, $z = 6$ implies that the triple must be one of $(3, 4, 6)$ and $(2, 5, 6)$. But B has deduced that x is 1 or 2, and so if B turned $z = 6$, she would know the numbers are $(2, 5, 6)$. Hence $z \neq 6$.

Moreover, if $z \leq 5$, then $S \leq 3 + 4 + 5 = 12$, and so z can only be 7 or 8. Recalling x is either 1 or 2, then everybody could figure out that there are only the four possibilities

$$(x, y, z) = (1, y, 7), (1, y, 8), (2, y, 7), \text{ and } (2, y, 8).$$

Hence if $y = 3$, the triple would have to be $(2, 3, 8)$, for none of the others yield a total of 13. Also, if $y = 5$, only $(1, 5, 7)$ is feasible. Hence it must be that $y = 4$, which leaves C wondering whether it's $(1, 4, 8)$ or $(2, 4, 7)$.

4. (From the 1988 Chinese Training Camp)

Mr. A thinks of a 2-digit positive integer n, the **sum** of whose digits he divulges privately to Mr. B, and the **number** of positive divisors of which he confides to Mr. C. Furthermore, he informs them of the character of the number that each has been told.

Running into each other in the street one day, B declared: "I don't know what A's number is," to which C replied: "Neither do I, but I know whether it's odd or even." Then B said: "Really! In that case, I know what the number is," after which C declared: "You do? Then so do I."

What number was A thinking of?

Just like the AHSME problem, this is just a matter of grinding away at the possibilities until there is only one left, and such a prospect is often less than inspiring. However, it's fascinating that it is actually possible to figure out A's number from such scanty information and, as it turns out, a certain amount of respectable mathematics and clear thinking is involved.

(a) Let's denote by b and c the numbers told to B and C, respectively. We begin by showing that c cannot exceed 12.

Recall that, if the prime decomposition of n is $p_1^i p_2^j \cdots p_t^k$, then the number of positive divisors of n is

$$\tau(n) = (i+1)(j+1)\cdots(k+1).$$

Now, since $2 \cdot 3 \cdot 5 \cdot 7 = 210$ and $2^7 = 128$, no 2-digit number has more than three different prime divisors or an exponent greater than 6. In fact, there are only five 2-digit numbers with three different prime divisors:

$$2 \cdot 3 \cdot 5 = 30, \quad 2 \cdot 3 \cdot 7 = 42, \quad 2 \cdot 3 \cdot 11 = 66,$$
$$2 \cdot 5 \cdot 7 = 70, \quad 2 \cdot 3 \cdot 13 = 78,$$

for each of which, then, $\tau = 2 \cdot 2 \cdot 2 = 8$.

Also, there are only three such numbers with exponents bigger than 4:

$$2^6 = 64, \quad 2^5 = 32, \quad 2^5 \cdot 3 = 96,$$

for which $\tau = 7, 6$, and $6 \cdot 2 = 12$, respectively.

Even the exponent 4 occurs only in the four numbers

$$2^4 = 16, \quad 2^4 \cdot 3 = 48, \quad 2^4 \cdot 5 = 80, \quad 3^4 = 81,$$

giving $\tau(16) = \tau(81) = 5$ and $\tau(48) = \tau(80) = 10$.

The remaining numbers, then, are of the form $p_1^i p_2^j$, where $i, j \leq 3$, for which $\tau \leq 4 \cdot 4 = 16$. Observing that $2^3 3^3 = 216$, it follows that no remaining number has a value of τ exceeding $4 \cdot 3 = 12$. Hence $\tau \leq 12$ for all 2-digit numbers.

Clearly, c couldn't be the prime 11, since $\tau(n) = 11$ requires $n = p^{10} \geq 2^{10}$. Therefore c can only be one of the ten values $\{2, 3, 4, 5, 6, 7, 8, 9, 10, 12\}$, five of which can be eliminated without much effort as follows.

(b) Recall that $\tau(n)$ is odd if and only if n is a perfect square. Thus, let us begin with the odd values. Observe that, since 2^7 is a three-digit number, n cannot be a prime power p^a with $a > 6$.

$c = 3$: then $n = p^2$, and since $2^2, 3^2$, and 11^2 are not 2-digit numbers, n could only be $5^2 = 25$ or $7^2 = 49$. In this

Nine Miscellaneous Problems

case C **would** know that n is odd, and so $c = 3$ is a possibility.

$c = 5$: then $n = p^4$, making n either $2^4 = 16$ or $3^4 = 81$, and C would **not** know whether n is even or odd. Hence c cannot be 5.

$c = 7$: then $n = p^6 = 2^6 = 64$, which would disclose n to C straightaway. But C does not know n to begin with; hence $c \neq 7$.

$c = 9$: then $n = p_1^2 p_2^2$, and since $2^2 \cdot 5^2 = 100$, C would know that n can only be $2^2 \cdot 3^2 = 36$. Hence $c \neq 9$.

$c = 2$: n would be a 2-digit prime p, and therefore odd; hence $c = 2$ is a possibility.

$c = 4$: then $n = p_1 p_2$ or $p^3 \in \{10, 14, 15, 21, 22, \ldots\}$, which does **not** inform C of the parity of n; hence $c \neq 4$.

$c = 6$: then $n = p^5$ or $p_1^2 p_2 \in \{32, 12, 18, 20, 45, \ldots\}$, again denying C the parity of n; hence $c \neq 6$.

$c = 8$: then $n = p_1^3 p_2$ or $p_1 p_2 p_3 \in \{24, 30, 40, 42, 54, 56, 60, 70, 78, 88\}$, all of which are even numbers; hence $c = 8$ is a possibility.

$c = 10$: then $n = p_1^4 p_2 \in \{48, 80\}$, giving another possibility.

$c = 12$: then $n = p_1^5 p_2$ or $p_1^3 p_2^2 \in \{60, 72, 84, 90, 96\}$, a final possibility.

Therefore c can only be one of the numbers $\{2, 3, 8, 10, 12\}$.

(c) Next we need to work out the corresponding values of n and b. This is a little tedious but it puts us in the homestretch with the following table.

c	n	b
2	11, 13, 17, 19, 23, 29, 31, 37, 41, 43, 47, 53, 59, 61, 67, 71, 73, 79, 83, 89, 97	**2**, 4, 8, 10, 5, 11, 4, 10, 5, 7, 11, 8, **14**, 7, 13, 8, 10, 16, 11, **17**, 16
3	25, 49	7, 13
8	24, 30, 40, 42, 54, 56, 60, 70, 78, 88	6, **3**, 4, 6, 9, 11, 6, 7, 15, 16
10	48, 80	12, 8
12	60, 72, 84, 90, 96	6, 9, 13, 9, 15

The values of n and b are listed in corresponding order; in the top line, for instance, $n = 11, 13, 17, \ldots$ pair, respectively, with $b = 2, 4, 8, \ldots$.

Now, both B and C can work out this table just as we have. B would know that things are restricted to these values as soon as he hears that C can tell whether n is odd or even (granted B would have to be an amazing character to work this out in his head while carrying on a conversation on a street corner). However, if b were to appear more than once among these possibilities, B would not be able to pinpoint the value of n:

for example, if $b = 4$, then B would not be able to decide between $n = 13, 31$, and 40 which occur among the possibilities for $c = 2$ and $c = 8$.

But B **is** able to determine n, and so his number must appear in the table **only once**, reducing the possible values of b to the four numbers $\{2, 14, 17, 3\}$, all of which are associated with the values $c = 2$ and $c = 8$. Whichever of these four values b might be, it is sufficient to settle things for B; for example

if b were 17, then B would know that c must be 2 and $n = 89$.

Now, C knows the value of c, and he also knows that, with this table, B is able to determine n. Hence if c were 2, C could see from the table that b would have to be one of the three numbers 2, 14, 17, but he wouldn't know which one, and hence C would **not** be able to determine n. But, at this point, C **is** able to settle the matter, and so c must be 8 rather than 2, in which case C observes that, since b must occur in the table just once, the only possibility is $b = 3$ and $n = 30$.

Comment In hindsight, we can see that B wouldn't have to figure out a great many things. With $b = 3$, he would know from the beginning that n could only be 12, 21, or 30, for which $c = 6, 4$, and 8, respectively. If he then noticed that each of $c = 6$ and $c = 4$ is a characteristic of both odd and even 2-digit numbers, he would immediately know, upon hearing that C knows the parity of n, that it couldn't be either 12 or 21, leaving $n = 30$.

5. (On The Median Triangle)

Prove that the area of the triangle constructed from the medians of $\triangle ABC$ is $\frac{3}{4} \triangle ABC$.

Let the medians be *AD*, *BE*, and *CF*, the centroid *G*, and *K* the midpoint of *GC* (Figure 3). Also let the area of △*EGK* be *x*.

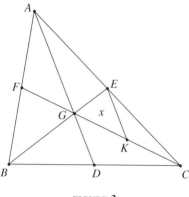

FIGURE 3

Since the medians trisect one another, then

$$GK = \tfrac{1}{2}GC = \tfrac{1}{2} \cdot \tfrac{2}{3}CF = \tfrac{1}{3} \text{ median } CF.$$

Also,

$$GE = \tfrac{1}{3} \text{ median } BE,$$

and since *KE* joins the midpoints of two sides of △*AGC*,

$$KE = \tfrac{1}{2}AG = \tfrac{1}{2} \cdot \tfrac{2}{3}AD = \tfrac{1}{3} \text{ median } AD.$$

Thus △*GEK* has sides that are each $\tfrac{1}{3}$ of the sides of the "median" triangle in question. Thus △*GEK* is similar to the median triangle and one-ninth its area. Thus the median triangle has area 9*x* and it remains only to show that △*ABC* = 12*x*.

Now, median *EK* bisects △*EGC*, and so △*EGC* = 2*x*.

Also, *BG* : *GE* = 2 : 1, and so △*BGC* = 2 · △*EGC* = 4*x*. Thus △*BEC* = 6*x*, and since *BE* is a median, △*BEC* is half of △*ABC*, and the conclusion follows.

6. (A Construction Problem of Howard Eves)

The following problem was proposed by Howard Eves in the $\pi\mu\varepsilon$ Journal, Vol. 10, Fall, 1996:

Determine a straightedge and compass construction for a triangle ABC given the lengths a and b of its sides BC and AC and the fact that the medians to these sides are perpendicular (Figure 4).

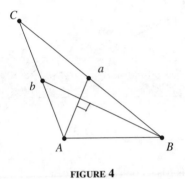

FIGURE 4

The wonderful solution by Ben Smith (Miami University, Oxford, Ohio) appeared in the Fall 1997 issue.

First lay down a segment BC of length a (Figure 5), and find its midpoint A'. Then construct the midpoint X of $A'C$ and draw the circle with diameter BX. With center C and radius $\frac{1}{2}b$, mark B' on the circle and extend CB' its own length to A. Then $BC = a$ and $AC = b$, and AA' and BB' are the medians to these sides in $\triangle ABC$.

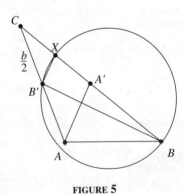

FIGURE 5

Now, B' and X are the midpoints of two sides of $\triangle CAA'$, and so $B'X$ and AA' are parallel. But $B'X$ and BB' are perpendicular since $\angle BB'X$ is an angle in a semicircle. Hence AA' and BB' are perpendicular, and the solution is complete.

7. (A Variation on a Problem of Regiomontanus)

In 1471 Regiomontanus asked how far from the base of a statue one should stand in order for the statue to appear biggest. Lorch's clever solution is given in my *Ingenuity in Mathematics* (The Anneli Lax New Mathematical Library Series, MAA, vol. 23, 1970).

In the problems section of the March 1998 issue of *The College Mathematics Journal*, (problem 624, page 153), Harry Sedinger of St. Bonaventure University in New York proposed what is essentially the following variant:

Let C be an interior point in the segment AB, and let Z be a variable point on a ray r through B (Figure 6). Suppose r makes an angle θ with AB, $0 < \theta < \pi$. Clearly the angle $\varphi = \angle AZC$ changes continuously as Z varies on r, and since φ is very small when Z is close to B, increases with BZ for awhile, and then decreases as BZ gets large, the angle attains a maximum value at some point P on r.

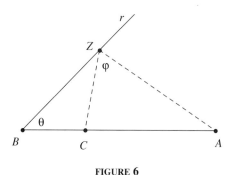

FIGURE 6

Prove the remarkable result that BP is always the same length no matter what the value of θ.

The key is to observe that the circle through A and C which is tangent to r touches it at the point P (Figure 7):

clearly AC subtends a smaller angle at every point outside the circle than it does at P on the circle.

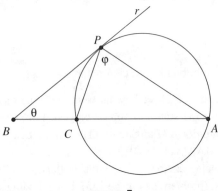

FIGURE 7

And since BP is a tangent, we have

$$BP^2 = BA \cdot BC,$$

which is the same for all θ.

(We might observe that the points P lie on the circle with center B that is orthogonal to the family of circles through A and C.)

8. (From the 1994 Mathematical Competition of Taiwan)

What is the area of the largest regular hexagon H that can be inscribed in (or covered by) a unit square S?

It is a pleasure to thank Professor Liang-shin Hahn (University of New Mexico) for kindly sending me his English translation of this interesting competition. This problem also appears as Problem 46 in *Which Way Did The Bicycle Go?* (Dolciani Series, Vol. 18), where it is credited to Cleon Richmeyer.

This problem did not occur on its own but as part (c) of question 5, whose first two parts contained helpful pointers. In our discussion, however, we will take an independent approach.

I. There doesn't seem to be any way of dealing with this problem without assuming that H and S are concentric and that H makes contact with each side of S in exactly one point. It is easy to assume these fundamentals, which I discovered aren't simple preliminaries that can be settled with a sweep of the hand. I had to work hard to prove them and my arguments weren't very exciting. After all, then, let us take the easy way out and simply take these things for granted.

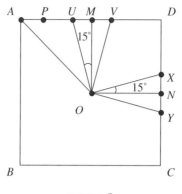

FIGURE 8

II. Let S be the square $ABCD$, with center O, and let M and N be the midpoints, respectively, of AD and CD (Figure 8). Clearly, a single point of contact would have to be a **vertex** of H. Let an angle of 15° be drawn on each side of OM and ON to give 30°-sectors UOV and XOY. We shall show that a vertex of contact P on AD must lie in the closed interval UV.

P would have to occur either in AM or MD; for definiteness, suppose it occurs in AM, possibly at M. Suppose also that it doesn't lie in UM, but outside in the segment AU (Figure 8).

Now, angles AOX and UOY are easily calculated to be 120°, which is the very angle that two consecutive sides of H subtend at its center, which we are assuming also to be the center O of the square. Thus if a vertex P lies in $\angle AOU$, then the second vertex Q around H from P lies in $\angle XOY$. That is, if P fails to lie in UM, then a vertex Q will lie in $\angle XOY$. Now, it is the same radial distance r from O to each vertex of H, and so $OP = OQ$. But since either P or Q lies in one of the central sections UM or XY, r cannot exceed OU $(= OX = OY)$. Thus $OP \leq OU$ and it follows that P must lie in the closed interval UM.

Now, each side of a regular hexagon subtends an angle of 60° at its center and, with the center, each side determines an **equilateral** triangle whose area is one-sixth the area of the hexagon. Thus a side of H cannot exceed the distance r from O to a vertex, and since $r \leq OU$, it remains to determine whether H can have a side as big as OU.

It is not difficult to see that it can. By symmetry, the arms of 60° angles OUK and OYK meet on diagonal OD (Figure 9). Since $\angle UOK = 15° + 45° = 60°$, $\triangle UOK$, and similarly $\triangle KOY$, is equilateral, and a half-turn about O completes an inscribed regular hexagon $UKYU'K'Y'$: clearly

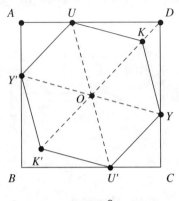

FIGURE 9

the six segments from O to the vertices are all equal in length and they determine a fan of $60°$ angles at O. Having the maximum radial distance $r = OU$, it follows that this hexagon is of maximum area.

III. The required area is therefore

$$6(\triangle UOK) = 6\left(\frac{1}{2} \cdot OU \cdot OK \cdot \sin 60°\right)$$
$$= \frac{3\sqrt{3} \cdot OU^2}{2},$$

and it remains only to calculate OU^2. From right triangle UOM in Figure 9 we have $OM = \frac{1}{2}$, giving $UO = \frac{1}{2}\sec 15°$, and hence $UO^2 = \frac{1}{4}\sec^2 15°$.

From

$$\frac{\sqrt{3}}{2} = \cos 30° = 2\cos^2 15° - 1,$$

we obtain

$$\frac{\sqrt{3}+2}{4} = \cos^2 15°$$

and hence

$$\frac{1}{4}\sec^2 15° = \frac{1}{4} \cdot \frac{4}{2+\sqrt{3}} = \frac{1}{2+\sqrt{3}} \cdot \frac{2-\sqrt{3}}{2-\sqrt{3}} = 2-\sqrt{3}.$$

Thus the required area is

$$\frac{3\sqrt{3} \cdot (2-\sqrt{3})}{2} = \frac{6\sqrt{3}-9}{2} = 0.696\ldots.$$

9. (Two Problems From the Second Selection Test for the 1991 Japanese International Olympiad Team)

(i) Prove that, in any 16-digit positive integer, there is some consecutive string of digits whose product is a perfect square.

(Solution due to Larry Rice, University of Toronto Schools.)
If any digit of a 16-digit integer $n = a_1 a_2 \ldots a_{16}$ is itself a square, i.e., 0, 1, 4, or 9, the conclusion follows trivially. We need to consider, then, only integers whose digits are 2, 3, 5, 6, 7, and 8. Thus every product under consideration is of the form $2^a 3^b 5^c 7^d$, and we would like to show that, for some product, the exponents a, b, c, d are all even.

Now, there are 16 products that begin with a_1, namely $a_1, a_1 a_2, a_1 a_2 a_3, \ldots$, and there are 16 parity classes for (a, b, c, d): (odd, odd, odd, odd), (odd, odd, odd, even), ..., (even, even, even, even). If (a, b, c, d) is (even, even, even, even) for one of these products, the conclusion follows. If it is not, these 16 products fall into the other 15 classes, and the pigeonhole principle implies some two of them, $a_1 \ldots a_k$ and $a_1 \ldots a_t$, where $t > k$, lie in the same class. Since the difference between two odd exponents or two even exponents is even, it follows that the quotient $(a_1 \ldots a_t)/(a_1 \ldots a_k) = a_{k+1} a_{k+2} \ldots a_t$ is a perfect square.

(ii) Let S be a set of n distinct points in the plane, where $n \geq 2$. Show that there are two distinct points P_i and P_j of S such that the circle on diameter $P_i P_j$ contains in its interior and on its circumference a total of at least $[n/3]$ distinct points of S, where $[m]$ is the greatest integer $\leq m$.

Let C be the circle of minimum radius that encloses the set S: one can imagine a large circle around S being shrunk until a tight contact is made with the outlying points of S. There are only two kinds of contact that can result:

(i) some two points P_i and P_j of S lie at the ends of a diameter of C, thus preventing further shrinking (Figure 10, Case (i)),
(ii) some three points of contact, P_i, P_j, P_k, determine an **acute**-angled triangle (Figure 10, Case (ii)).

Note that an obtuse-angled triangle, on its own, is not enough to put an end to the shrinking. The three vertices of an obtuse-angled triangle must be contained in the interior of a semicircular arc (Figure 10, Case (iii)), implying that

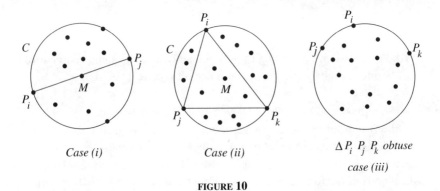

FIGURE 10

more than a semicircular arc does **not** make contact with S, in which case the circle can be pushed away from the points of contact and shrunk further.

In case (i), then, the circle on diameter $P_i P_j$ contains the entire set S.

In case (ii), the three circles whose diameters are the sides of $\triangle P_i P_j P_k$ contain the whole set S, in which case at least one of these circles must contain at least $[n/3]$ points of S:

since the angle in a semicircle is a right angle, any two of these circles contain the triangle itself (Figure 11), and it is clear that each circle covers the adjacent segment of the underlying circumcircle of S.

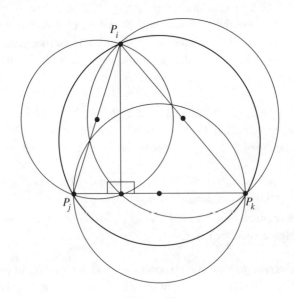

FIGURE 11

SECTION 12
A Problem in Coin-Tossing*

In the mid-1960s, my colleague Ron Read cracked the following problem which had gone unsolved since 1941. His clever solution appeared in the *American Mathematical Monthly* (1966, pages 177–179) in a short note which is distinguished by his clear, smooth style of writing.

"Three men play the old coin-matching game "odd man wins." As usual, in the case of a tie, they just toss again. If the stakes of the players consist of x, y, and z coins, what is the average (or expected) number of tosses that will occur before one of them goes broke?"

Let the players be labelled "first," "second," and "third," and let $E(x, y, z)$ denote the expected number of tosses when the first man has x coins, the second y coins, and the third z coins. Now, the first toss of the game must be one of the eight possible triples, which may be displayed in pairs as follows:

(i)   (HHH), (TTT),   (ii)   (HTT), (THH),
(iii)   (HTH), (THT),   (iv)   (HHT), (TTH).

A toss yields a triple in pair (i) one-quarter of the time; similarly for the other pairs. In case (i), nobody wins, and $E(x, y, z)$ further tosses are expected in order to finish the game. Thus, one-quarter of the time the total number of tosses expected is

$$1 + E(x, y, z).$$

If a triple in (ii) occurs on the first toss, the first man wins, and the number of tosses expected to complete the game is $E(x + 2, y - 1, z - 1)$. Thus, one-quarter of the time, the total number of tosses expected is

$$1 + E(x + 2, y - 1, z - 1).$$

---

*This is a reworked version of material from my "Mathematical Gems" column in the *Two-Year College Mathematics Journal*, Vol. 10, March, 1979.

Handling (iii) and (iv) in like fashion, we have that, over four games, the expected number of tosses is

$$4E(x, y, z) = \big[1 + E(x, y, z)\big] + \big[1 + E(x+2, y-1, z-1)\big]$$
$$+ \big[1 + E(x-1, y+2, z-1)\big]$$
$$+ \big[1 + E(x-1, y-1, z+2)\big]$$
$$= 4 + E(x, y, z) + E(x+2, y-1, z-1)$$
$$+ E(x-1, y+2, z-1) + E(x-1, y-1, z+2),$$

giving

$$E(x, y, z) = \tfrac{4}{3} + \tfrac{1}{3}E(x+2, y-1, z-1) + \tfrac{1}{3}E(x-1, y+2, z-1)$$
$$+ \tfrac{1}{3}E(x-1, y-1, z+2).$$

At this point it is convenient to change notation slightly to

$$x = p+1, \quad y = q+1, \quad z = r+1.$$

In these terms, we have

$$E(p+1, q+1, r+1) = \tfrac{4}{3} + \tfrac{1}{3}E(p+3, q, r) + \tfrac{1}{3}E(p, q+3, r)$$
$$+ \tfrac{1}{3}E(p, q, r+3). \tag{1}$$

Now, the total amount of money in the game never changes, and so the sum of the arguments is always the constant

$$N = x + y + z = p + 1 + q + 1 + r + 1.$$

Let us illustrate the use of the recursion (1) in the case of $N = 7$.

Clearly $E(a, b, c) = 0$ when any of a, b, c is 0 (the game is over). Thus, for $p = 4, q = r = 0$, we get

$$E(5, 1, 1) = \tfrac{4}{3}.$$

Assigning the triples $(3, 1, 0)$, $(2, 2, 0)$, $(2, 1, 1)$ to (p, q, r) in (1), we obtain the set of three equations

$$E(4, 2, 1) = \tfrac{4}{3} + \tfrac{1}{3}E(3, 1, 3),$$
$$E(3, 3, 1) = \tfrac{4}{3} + \tfrac{1}{3}E(2, 2, 3),$$

and

$$E(3, 2, 2) = \tfrac{4}{3} + \tfrac{1}{3}E(5, 1, 1) + \tfrac{1}{3}E(2, 4, 1) + \tfrac{1}{3}E(2, 1, 4).$$

Now, if the stakes are 4, 2, and 1 coins, then, regardless of who has which amount, it is clear that the expected number of tosses to complete the game will be the same; it's only the unordered triple of the holdings that counts. Hence

$$E(4, 2, 1) = E(2, 4, 1) = E(2, 1, 4),$$

and similarly,

$$E(3, 1, 3) = E(3, 3, 1),$$

and

$$E(2, 2, 3) = E(3, 2, 2).$$

Since we already found $E(5, 1, 1) = \frac{4}{3}$, this set of three equations contains only the three unknowns $E(4, 2, 1)$, $E(3, 3, 1)$, and $E(3, 2, 2)$, and they can be solved to give, with $E(5, 1, 1)$, the whole story for $N = 7$ (7 can be partitioned into unordered triples of positive integers in only these four ways).

Now this might be considered a satisfactory solution for small values of N like 7, but it is not very satisfying to leave the general values of $E(x, y, z)$ tied up in a set of n equations in n unknowns. We want a general formula that gives $E(x, y, z)$ explicitly in terms of x, y, and z.

It often turns out that the key to a difficult problem is to be found in the solution of a simpler, related problem. Let us look, then, at the popular two-man version of this game, which awards the win to "heads" when the coins come up different and has the toss repeated when they are the same.

An analysis which parallels that used above easily yields

(i)   (HH): $1 + E(x, y)$ tosses;
(ii) (TT): $1 + E(x, y)$ tosses;
(iii) (HT): $1 + E(x + 1, y - 1)$ tosses;
(iv) (TH): $1 + E(x - 1, y + 1)$ tosses,

$$4E(x, y) = 2\big[1 + E(x, y)\big] + \big[1 + E(x + 1, y - 1)\big]$$
$$+ \big[1 + E(x - 1, y + 1)\big],$$
$$E(x, y) = 2 + \tfrac{1}{2}E(x + 1, y - 1) + \tfrac{1}{2}E(x - 1, y + 1),$$

which, in terms of $x = p + 1$, $y = q + 1$, is

$$E(p + 1, q + 1) = 2 + \tfrac{1}{2}E(p + 2, q) + \tfrac{1}{2}E(p, q + 2).$$

Now, $x + y = (p + 1) + (q + 1) = N$ is constant, implying that there is only one independent variable in this function E; the values are all of the form $E(a, N - a)$, which we shall denote by u_a. Accordingly, the recursion is

$$u_{p+1} = 2 + \tfrac{1}{2}u_{p+2} + \tfrac{1}{2}u_p.$$

that is,

$$u_{p+2} - 2u_{p+1} + u_p = -4.$$

The standard solution of this difference equation gives

$$u_p = Cp + D - 2p(p - 1),$$

where C and D are constants to be determined (a very readable account of difference equations is given in the book *Difference Equations* by Samuel Goldberg, John Wiley and Sons, 1958). Because $u_0 = u_N = 0$, we get

$$D = 0 \quad \text{and} \quad 0 = CN - 2N(N - 1),$$

which gives $C = 2(N - 1)$. Thus

$$u_p = 2(N - 1)p - 2p(p - 1) = 2p(N - p).$$

That is to say,

$$E(p, N - p) = 2p(N - p),$$

which is twice the product of the arguments, and we have in general that

$$E(x, y) = 2xy.$$

It's a lot to hope for, but it might be possible that the three-man game also has a solution

$$E(x, y, z) = Kxyz \quad \text{for some constant } K.$$

Substituting this in the recursion (1), which we recall is

$$E(p + 1, q + 1, r + 1) = \tfrac{4}{3} + \tfrac{1}{3}E(p + 3, q, r)$$
$$+ \tfrac{1}{3}E(p, q + 3, r) + \tfrac{1}{3}E(p, q, r + 3),$$

where $E(x, y, z) = 0$ for any of x, y, z, equal to zero, we get

$$K(p + 1)(q + 1)(r + 1) = \tfrac{4}{3} + \tfrac{1}{3}K(p + 3)qr$$
$$+ \tfrac{1}{3}Kp(q + 3)r + \tfrac{1}{3}Kpq(r + 3).$$

Solving for K, we get

$$K = \frac{4}{3(p+q+r+1)},$$

which is indeed constant, and since $p+q+r+1 = x+y+z-2$, we obtain the general formula

$$E(x, y, z) = \frac{4xyz}{3(x+y+z-2)},$$

that is,

$$E(x, y, z) = \frac{4xyz}{3(N-2)}.$$

Comment This result is one solution of recursion (1). Since recursions can have more than one solution, before we can claim to have solved the original problem we need to show that our result is the only solution of the recursion. This can be established by a standard argument in the theory of absorbing Markov chains. We shall not carry the matter further in this essay, but the interested reader will find a discussion of Markov chains in the famous book *Probability* by W. Feller (McGraw Hill, 1960).

SECTION 13
Semi-regular Lattice Polygons*

The point (x, y) in a coordinate plane is a "lattice point" if both x and y are integers, and an n-gon having its vertices at lattice points is a "lattice polygon." It has been known for at least half a century that a lattice polygon can be *regular* only if $n = 4$, i.e., the only regular lattice polygon is a square. (This is proved in *Combinatorial Geometry in the Plane*, by Hadwiger, Debrunner, and Klee, published by Holt, Rinehart, and Winston, 1964. See page 4). Of course, regularity requires both equal sides and equal angles. A polygon that enjoys only one of these properties is called semi-regular. It is our purpose to determine the values of n for which semi-regular lattice polygons exist. The theorems below, and their ingenious proofs, are due to Dean Hoffman of Auburn University, Alabama.

Theorem 1. *There exists an* equilateral *lattice polygon if and only if n is an even number ≥ 4.*

Theorem 2. *There exists an* equiangular *lattice polygon if and only if $n = 4$ or $n = 8$.*

Proof of Theorem 1:

(a) Suppose $n \geq 4$ is even.
It is convenient to think of (x, y) both as a lattice point and also as the vector from the origin to the point (x, y). Now, because of the regular spacing of the lattice points in the plane, if a vector v begins and ends at a lattice point, then, no matter to which lattice point its initial point may be transferred, without changing the length or direction of the vector, its terminal point will also be at a lattice point (Figure 1).

---
*This is a reworked version of material from my "Mathematical Gems" column in the *Two-Year College Mathematics Journal*, Vol. 13, January, 1982.

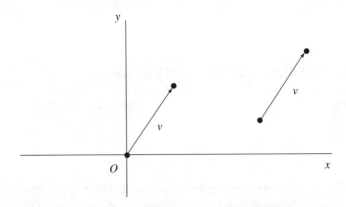

FIGURE 1

Clearly any square in the lattice takes care of the case $n = 4$. To see that there is an equilateral lattice polygon for every even number $n \geq 6$, consider the lattice vector that joins the origin to the point $(3, 4)$, and the vector from the origin to $(-3, 4)$. The length of each of these vectors is 5, and stringing together two copies of each with two copies of the vector that joins the origin to the point $(5, 0)$, a "zigzag" equilateral lattice hexagon may be constructed as shown in Figure 2. Since such a polygon of any height may be constructed by zigzagging farther up the y-axis, it follows that an equilateral lattice n-gon exists for every even number $n \geq 6$.

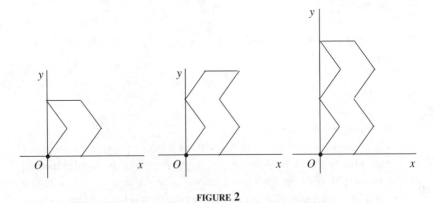

FIGURE 2

(b) Suppose n is odd.

We shall see that it is impossible for an equilateral lattice polygon to have an odd number of sides. Proceeding indirectly, suppose n is odd and the lattice n-gon P_n is equilateral with sides of length d.

Clearly the run and rise between two lattice points are integers. Thus if $A(x_1, y_1)$ and $B(x_2, y_2)$ are consecutive vertices of P_n, we have

$$d^2 = (x_2 - x_1)^2 + (y_2 - y_1)^2, \quad \text{an integer.}$$

Next let us consider a neat argument that shows each of $x_2 - x_1$ and $y_2 - y_1$ is an *even* integer.

(i) First we show that d^2 is even.

To this end, let each lattice point (x, y) in the plane be colored blue if x and y have the same parity, and colored red if x and y have opposite parity. Now, as one walks around P_n from one vertex to the next, the vertices could alternate in color all the way around the polygon only if the number of vertices is even. Since this is not the case, some edge of the polygon must have vertices $A(x_1, y_1)$ and $B(x_2, y_2)$ which are the same color. Whether they are both blue or both red, it follows that $x_2 - x_1$ and $y_2 - y_1$ have the same parity, implying d^2 is even:

if A and B are both blue, then each of them is either (even, even) or (odd, odd), giving rise to four cases.

For the case A(odd, odd) and B(even, even), for example,

$$x_2 - x_1 \quad \text{and} \quad y_2 - y_1 \quad \text{are both (even–odd)} = \text{odd.}$$

The other three cases for A and B both blue, and the four cases when they are both red, are similar.

(ii) d^2 is divisible by 4.

Although d^2 is an integer, we don't know that d itself is an integer. Thus we cannot immediately conclude that d^2 is divisible by 4: presumably it might be a number like 34, of the form $4k + 2$. In order to show that d^2 is, in fact, divisible by 4, we turn to a new coloring of the lattice points.

Accordingly, let us dispense with the present coloring and recolor the lattice points in columns as follows. Let the entire column of lattice points (x, y) be colored blue if x is even and red if x is odd. Thus the vertical lattice lines alternate blue and red all across the plane.

It is still true that some edge ST of P_n must have both its vertices the same color (otherwise n would have to be even). If S and T are blue, then the run $x_2 - x_1$ between them would be (even–even), and if red, (odd–

odd), and so this coloring scheme implies that $x_2 - x_1$ is even in any case. But what about the rise $y_2 - y_1$? Since d^2 and $x_2 - x_1$ are both even, and furthermore, since

$$d^2 = (x_2 - x_1)^2 + (y_2 - y_1)^2,$$

it follows that $(y_2 - y_1)^2$ must be even, and hence also $y_2 - y_1$. Thus for some even integers $2p$ and $2q$ we have

$$d^2 = (2p)^2 + (2q)^2 = 4(p^2 + q^2) \equiv 0 \pmod{4}.$$

(iii) $x_2 - x_1$ and $y_2 - y_1$ are both even for every edge AB of P_n.

From this special edge ST, then, we have found that the run and rise between its ends are both even. But what about the other edges? Since P_n is equilateral, for every edge AB we have

$$AB^2 = \text{run}^2 + \text{rise}^2 = d^2 \equiv 0 \pmod{4}.$$

Unlike the case of d^2 even, this time we know that the run and rise are integers. Being the square of an integer, each of run^2 and rise^2 is $\equiv 0$ or to 1 (mod 4). In view of their sum being $\equiv 0 \pmod 4$, it follows that each of run^2 and rise^2 must be $\equiv 0 \pmod 4$. Hence the run and rise themselves must be even.

(iv) The Final Step.

Without loss of generality, suppose A is the origin of the lattice. Since all the rises and runs of the edges are even, it follows that the coordinates of each vertex of P_n are both even. Consequently, dividing all the coordinates by two results in an equilateral lattice polygon $P_n^{(1)}$ in which all sides have length $d_1 = \frac{1}{2}d$.

Now, just as in the case of P_n, $P_n^{(1)}$ is an equilateral lattice polygon with an odd number of sides, namely n, and so the coordinates of all its vertices must be even. Thus, dividing all the coordinates again by two, an equilateral lattice polygon $P_n^{(2)}$ of n sides is obtained whose side is of length $d_2 = \frac{1}{2}d_1 = \frac{1}{2^2}d$. Repeatedly dividing all the coordinates by two, we obtain a sequence of equilateral lattice polygons $P_n^{(m)}$ of n sides of lengths $d_m = \frac{d}{2^m}$. As m increases, 2^m eventually exceeds d, implying a lattice polygon of side $d_m < 1$, which is clearly impossible, and the indirect proof of Theorem 1 is complete.

I am grateful to James Tanton of The Math Circle for proposing the color schemes used in this proof. Dr. Tanton also observed that there exist color schemes that lead to the proof of Theorem 1 for a three-dimensional cubic lattice.

Proof of Theorem 2: There exists an *equiangular* lattice polygon if and only if

$$n = 4 \quad \text{or} \quad n = 8.$$

Suppose P is an equiangular lattice n-gon and that each angle of P is φ. Let the sides of P be oriented in the same cyclic direction and let the resulting vectors be translated to the origin (Figure 3). At the origin, the angle between consecutive sides z_1 and z_2 is not φ, but the constant exterior angle $\theta = \pi - \varphi$ which occurs at each vertex of P. Let the inclinations of z_1 and z_2 with the positive x-axis be α_1 and α_2, respectively. Now, since the endpoints of z_1 and z_2 are lattice points, the run and rise from the origin to these endpoints are integers, making $\tan \alpha_1$ and $\tan \alpha_2$ rational numbers. Therefore

$$\tan \theta = \tan(\alpha_2 - \alpha_1) = \frac{\tan \alpha_2 - \tan \alpha_1}{1 + \tan \alpha_2 \cdot \tan \alpha_1} = \text{a rational number}.$$

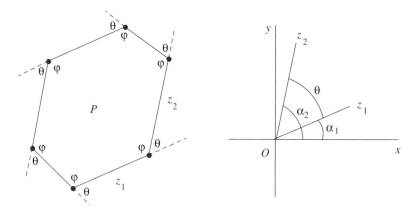

FIGURE 3

Traversing all the edges of P requires one to turn through the exterior angle θ at each vertex in order to continue along the next side. Altogether, then, one would turn through a total angle which is a whole number of revolutions— only one revolution if P is convex, but possibly more if P crosses over itself. Hence, for some positive integer k, we have the total $n\theta = 2\pi k$, and

$$\theta = \frac{2\pi k}{n}.$$

Letting $a = e^{i\theta}$, we have

$$a = e^{2\pi k i / n},$$
$$a^n = (e^{2\pi i})^k = 1,$$

and we also have

$$\left(\frac{1}{a}\right)^n = \frac{1}{a^n} = 1.$$

For this θ, then, the numbers a and $\frac{1}{a}$ are roots of $x^n - 1 = 0$.

A polynomial whose coefficients are integers and whose leading coefficient is unity is said to be a *monic* polynomial. A complex number that is a root of a monic polynomial equation is said to be an *algebraic integer*. Thus a and $\frac{1}{a}$, being roots of $x^n - 1 = 0$, are algebraic integers. Now, it is central to our proof that we know the number $a + \frac{1}{a}$ is an algebraic integer. Since it is known that the algebraic integers are closed under addition, the result is certainly true. However, instead of establishing it by going through the sophisticated proof of this general property of algebraic integers, let us follow a comfortable and instructive ad hoc approach.

Consider the *Chebyshev polynomials* $f_r(x)$, which are defined as follows:

$$f_0(x) = 2, \quad f_1(x) = x,$$

and for $r > 1$,

$$f_r(x) = x \cdot f_{r-1}(x) - f_{r-2}(x).$$

This generates a sequence of monic polynomials:

$$f_0(x) = 2, \quad f_1(x) = x, \quad f_2(x) = x^2 - 2, \quad f_3(x) = x^3 - 3x,$$
$$f_4(x) = x^4 - 4x^2 + 2, \ldots.$$

A very useful property of these polynomials is that, for all r,

$$f_r\left(a + \frac{1}{a}\right) = a^r + \frac{1}{a^r}.$$

We omit the proof since it is just a simple induction.

Recalling that $a^n = 1/a^n = 1$, we have

$$f_n\left(a + \frac{1}{a}\right) = a^n + \frac{1}{a^n} = 2.$$

Thus $a + \frac{1}{a}$ is a root of the monic polynomial equation

$$f_n(x) - 2 = 0,$$

implying

$$a + \frac{1}{a}, \quad \text{i.e., } e^{i\theta} + e^{-i\theta},$$

is an algebraic integer.

From the formula $\cos\theta = \frac{1}{2}(e^{i\theta} + e^{i\theta})$, it follows that $2\cos\theta$ is an algebraic integer. More to the point, we need to establish that its square, $t = 4\cos^2\theta$, is an algebraic integer. Fortunately it is easy to see that the square of an algebraic integer is also an algebraic integer. We demonstrate the standard approach with the following example.

Suppose x is known to be an algebraic integer because it satisfies the equation

$$x^8 - 4x^7 + 3x^6 + x^3 - 5 = 0,$$

and that we would like to prove that $y = x^2$ is also an algebraic integer. Transposing the odd powers to the right side and factoring out x, we get

$$x^8 + 3x^6 - 5 = x(4x^6 - x^2),$$

that is

$$y^4 + 3y^3 - 5 = x(4y^3 - y).$$

Squaring and substituting y for the single factor x^2 on the right side shows that y is indeed a root of a monic integral equation and is hence an algebraic integer. Thus we can count on $t = 4\cos^2\theta$ being an algebraic integer.

Moreover, it is easy to see that t is a rational number: because $\tan\theta$ is rational, we have that

$$1 + \tan^2\theta = \sec^2\theta = \frac{1}{\cos^2\theta} = \frac{4}{t}$$

is rational, from which it follows that t is rational.

However, we won't really be happy until we know that t is actually an *integer*. Fortunately, it is not difficult to show that an algebraic integer which is also rational is just an ordinary integer:

Suppose the rational number $x = \frac{p}{q}$, where q is positive and p and q are relatively prime, is an algebraic integer satisfying the integral polynomial equation

$$x^n + a_{n-1}x^{n-1} + \cdots + a_1 x + a_0 = 0.$$

Then, substituting $x = \frac{p}{q}$, we get

$$\left(\frac{p}{q}\right)n + a_{n-1}\left(\frac{p}{q}\right)^{n-1} + \cdots + a_1\left(\frac{p}{q}\right) + a_0 = 0,$$

and, multiplying by q^n, that

$$p^n + a_{n-1}p^{n-1}q + \cdots + a_1 pq^{n-1} + a_0 q^n = 0.$$

Then, since q divides all the other terms, q must also divide p^n. Since p and q are relatively prime, then q must be 1 and x an integer.

Finally, then, we have that $t = 4\cos^2\theta$ is an integer.

Since $|\cos^2\theta| \leq 1$, this restricts t to the five integers 0, 1, 2, 3, 4, which, in turn, confines $\cos^2\theta$ and θ to the values in the following table.

$t = 4\cos^2\theta$	0	1	2	3	4
$\cos^2\theta$	0	1/4	1/2	3/4	1
θ	$\pi/2$	$\pi/3$	$\pi/4$	$\pi/6$	0

Since $\theta = 0$ is obviously inadmissible, and $\tan\frac{\pi}{3}$ and $\tan\frac{\pi}{6}$ are not rational, the only feasible cases are $\theta = \frac{\pi}{2}$ and $\theta = \frac{\pi}{4}$, which correspond to $n = 4$ and $n = 8$, both of which can be realized (Figure 4).

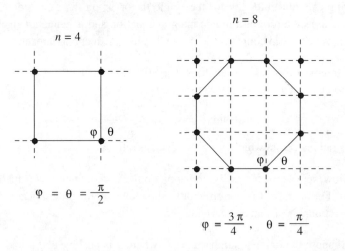

FIGURE 4

SECTION 14
Six Problems from the Canadian Open Mathematics Challenge

These six problems are from Canadian Open Mathematics Challenge contests, which are prepared for high school students by the University of Waterloo. They are followed by three exercises from sets of practice problems connected with these contests.

1. (From the 1996 contest; revised)

    $a_1, a_2, \ldots, a_{16}$ are positive real numbers whose sum is 100 and the sum of whose squares is 1000. Prove that none of the numbers is greater than 25.

    The key to this engaging result is a standard inequality called the Root-Mean-Square Inequality:

    the average of the squares of n nonnegative real numbers $\{a_1, a_2, \ldots, a_n\}$ is at least as great as the square of their average:

    $$\frac{1}{n}\sum a_i^2 \geq A^2 \quad \left(A = \frac{1}{n}\sum a_i\right).$$

The proof of this inequality was explicitly required in an earlier section of this problem and comprised the major part of the problem. The proof couldn't be easier once you think of letting $a_i = A + b_i$. Then

$$A = \frac{a_1 + a_2 + \cdots + a_n}{n}$$
$$= \frac{(A + b_1) + (A + b_2) + \cdots + (A + b_n)}{n}$$
$$= A + \frac{1}{n}\sum b_i,$$

revealing the interesting fact that $\sum b_i = 0$.

117

Thus it follows directly that

$$\frac{1}{n}\sum a_i^2 = \frac{1}{n}\sum (A^2 + 2Ab_i + b_i^2)$$
$$= \frac{1}{n}\left(nA^2 + 2A\sum b_i + \sum b_i^2\right)$$
$$= A^2 + \frac{1}{n}\sum b_i^2$$
$$\geq A^2.$$

Addressing the question at hand, suppose that one of the a_i were to exceed 25: let

$$a_{16} = 25 + x, \quad x > 0.$$

Then

$$a_1 + a_2 + \cdots + a_{15} + (25 + x) = 100$$

and

$$a_1^2 + a_2^2 + \cdots + a_{15}^2 + (625 + 50x + x^2) = 1000,$$

which reduce to

$$a_1 + a_2 + \cdots + a_{15} = 75 - x$$

and

$$a_1^2 + a_2^2 + \cdots + a_{15}^2 = 375 - 50x - x^2.$$

Applying the above inequality to $\{a_1, a_2, \ldots, a_{15}\}$, we have

$$\frac{1}{15}(375 - 50x - x^2) \geq \left(\frac{75-x}{15}\right)^2,$$
$$15(375 - 50x - x^2) \geq (75-x)^2,$$
$$5625 - 750x - 15x^2 \geq 5625 - 150x + x^2,$$
$$0 \geq 16x^2 + 600x,$$

and the contradiction

$$0 \geq x.$$

The conclusion follows.

Six Problems from the Canadian Open Mathematics Challenge 119

2. Five "bars" (i.e., closed intervals), each of length p, are to be placed on the number line to cover the infinite set of points $1, \frac{1}{2}, \frac{1}{3}, \ldots, \frac{1}{n}, \ldots$. What is the minimum length p which permits this to be accomplished?

Since $\frac{1}{n}$ approaches 0 as n grows indefinitely large, we can't avoid covering 0 since the bars are closed intervals. Thus the problem is not enlarged by also requiring 0 to be covered.

Clearly five bars of length $\frac{1}{5}$ will cover the whole unit interval $[0, 1]$ and so minimum $p \leq \frac{1}{5}$. Thus a bar of minimum length will not cover both 1 and $\frac{1}{2}$, whose distance apart is $\frac{1}{2}$.

We observe that p need not exceed $\frac{1}{9}$, for three bars of length $\frac{1}{9}$ can be placed over the intervals $[0, \frac{1}{9}]$, $[\frac{1}{9}, \frac{2}{9}]$, $[\frac{2}{9}, \frac{1}{3}]$ and one each over $\frac{1}{2}$ and 1.

Now, $\frac{1}{2}$ and $\frac{1}{3}$ are separated by a distance of $\frac{1}{6}$, which exceeds $\frac{1}{9}$, and so in a minimum covering, each of $\frac{1}{2}$ and $\frac{1}{3}$ must be covered by a different bar. Thus, in a minimum covering, the numbers 1 and $\frac{1}{2}$ each claim a bar of their own and the problem reduces to covering the interval $[0, \frac{1}{3}]$ with **three** bars.

Observing that in $[0, \frac{1}{3}]$ there are only two numbers bigger than $\frac{1}{5}$ which need to be covered, namely $\frac{1}{4}$ and $\frac{1}{3}$, even three bars of length $\frac{1}{10}$ suffice:

a single bar can cover both $\frac{1}{3}$ and $\frac{1}{4}$, and two more placed over the intervals $[0, \frac{1}{10}]$ and $[\frac{1}{10}, \frac{1}{5}]$ cover the rest.

Hence minimum $p \leq \frac{1}{10}$, and it remains to investigate whether any improvement can be made on $p = \frac{1}{10}$.

If $p < \frac{1}{10}$, then, for the numbers $\{0, \frac{1}{10}, \frac{1}{5}, \frac{1}{3}\}$, in which each is at least one-tenth of a unit from its nearest neighbor in the set, each would need a bar of its own, for an unacceptable total of four bars.

Hence p cannot be less than $\frac{1}{10}$, and the minimum p is indeed $\frac{1}{10}$.

3. (From the 1996 contest; revised)

Find all the lattice points (x, y) which lie on the graph of

$$6x^2 - 3xy - 13x + 5y + 11 = 0.$$

Suppose that (x, y) is a lattice point on the curve. While there is little hope of getting rid of the quadratic terms $6x^2 - 3xy$, at least we can consolidate them into a single term with the substitution $y = 2x - k$, where k is an appropriate integer. In this case, we obtain

$$0 = 6x^2 - 3xy - 13x + 5y + 11$$
$$= 6x^2 - 3x(2x - k) - 13x + 5(2x - k) + 11$$
$$= (3k - 3)x - 5k + 11,$$

giving

$$(3k - 3)x = 5k - 11.$$

Thus $5k - 11$ must be divisible by 3, which implies $k \equiv 1 \pmod{3}$, that is,

$$k = 3t + 1 \quad \text{for some integer } t.$$

Substituting in the equation, we get

$$9tx = 15t - 6,$$
$$3tx = 5t - 2,$$

and

$$(5 - 3x)t = 2.$$

Being integers, the values of t and $5 - 3x$ are restricted to ± 1 and ± 2, implying the following table of possibilities (recall $k = 3t+1$ and $y = 2x-k$):

t:	1	2	−1	−2
$5 - 3x$:	2	1	−2	−1
x:	1	—	—	2
k:	4	7	−2	−5
y:	−2	—	—	9

Therefore the only two lattice points on the graph are $(1, -2)$ and $(2, 9)$.

4. Now for a nice geometry problem (from the 1996 contest)

In Figure 1, the length of diagonal BD in rectangle $ABCD$ is d and E is the foot of the perpendicular to BD from A. If E is one unit from BC and n units from CD, that is, if the lengths of perpendiculars EG and EF are 1 and n, respectively, prove that

$$d^{2/3} - n^{2/3} = 1.$$

Those fractional exponents threaten troublesome complications. Anyway, let's extend FE to meet AB at H, and let $BG = x$ and $DF = y$ (Figure 2). Then

Six Problems from the Canadian Open Mathematics Challenge

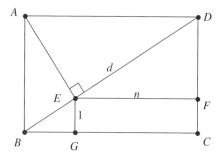

FIGURE 1

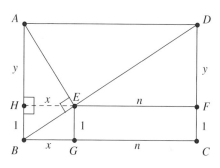

FIGURE 2

$$HE = x, \quad AH = y, \quad \text{and} \quad HB = 1.$$

Since EH is the altitude to the hypotenuse in right triangle ABE, we have

$$AH \cdot HB = HE^2, \quad \text{that is,} \quad y = x^2.$$

Also, from right triangle BEG,

$$BE = \sqrt{1 + x^2}.$$

Since EF is parallel to BC, $\triangle BEG$ and $\triangle DEF$ are similar, and because

$$\frac{DE}{BE} = \frac{DF}{EG} = \frac{y}{1} = x^2 \quad (\text{recall } y = x^2),$$

each side of $\triangle DEF$ is x^2 times the corresponding side of $\triangle BEG$. Hence

$$DE = x^2 \cdot BE = x^2\sqrt{1 + x^2},$$

and

$$n = EF = x^2 \cdot BG = x^3,$$

giving
$$n^{2/3} = x^2.$$

Also,
$$d = BD = BE + ED = \sqrt{1+x^2} + x^2\sqrt{1+x^2}$$
$$= (1+x^2)\sqrt{1+x^2} = (1+x^2)^{3/2},$$

and hence
$$d^{2/3} = 1 + x^2 = 1 + n^{2/3},$$

giving the desired
$$d^{2/3} - n^{2/3} = 1.$$

5. What is the greatest **integral** value taken by the function
$$\sqrt{x-174} + \sqrt{x+34}$$
for x an integer ≥ 174?

Suppose $\sqrt{x-174} + \sqrt{x+34} = k$, an integer. Then
$$\sqrt{x-174} = k - \sqrt{x+34}$$
and
$$x - 174 = k^2 - 2k\sqrt{x+34} + x + 34.$$

Solving for $\sqrt{x+34}$, we get a rational number, and since $x + 34$ is an integer, then $\sqrt{x+34}$ must be an integer, too. Similarly, $\sqrt{x-174}$ is also an integer. Let
$$x + 34 = a^2 \quad \text{and} \quad x - 174 = b^2.$$

Then
$$a^2 - b^2 = 208,$$
and
$$(a+b)(a-b) = 208.$$

Clearly a and b have the same parity, making both $a + b$ and $a - b$ even numbers. Thus the maximum value of $a + b$, which is the number we seek, is the greatest even divisor of 208 whose complementary divisor is also even, namely 104.

Six Problems from the Canadian Open Mathematics Challenge

6. If θ is an angle between 90° and 180°, and

$$\sin\theta + \cos\theta = \sqrt{2}/3,$$

what is the value of $\sin\theta - \cos\theta$?

Some problems have wonderfully straightforward solutions. Once you have thought of giving the desired expression a name and recalled that $\sin^2 A + \cos^2 A = 1$, the problem is practically solved.

Letting

$$\sin\theta - \cos\theta = x,$$

then, with

$$\sin\theta + \cos\theta = \frac{\sqrt{2}}{3},$$

we get

$$2\sin\theta = x + \frac{\sqrt{2}}{3}$$

and

$$2\cos\theta = \frac{\sqrt{2}}{3} - x.$$

Since

$$4\sin^2\theta + 4\cos^2\theta = 4,$$

then

$$\left(x + \frac{\sqrt{2}}{3}\right)^2 + \left(\frac{\sqrt{2}}{3} - x\right)^2 = 4,$$

$$2x^2 + \frac{4}{9} = 4,$$

$$2x^2 = \frac{32}{9}$$

and

$$x = \pm\frac{4}{3}.$$

Because θ is between 90° and 180°, $\sin\theta$ is positive and $\cos\theta$ is negative, making $(\sin\theta - \cos\theta)$ positive. Hence $x = \frac{4}{3}$.

Exercises

1. The six-digit integer *ababab* is equal to five times the product of three consecutive odd numbers. What odd numbers?

2. Angles A and B are both between 0 and 180°, and

   (i) $\sin A + \sin B = \sqrt{\dfrac{3}{2}}$, (ii) $\cos A + \cos B = \sqrt{\dfrac{1}{2}}$.

   What is $A + B$?

3. In the pyramid $PABCD$ in Figure 3, $PA = PB$ and $\angle APB = 20°$; also, $CP = CB$ and $\angle PCB = 100°$. Prove $AB + BC = AP$.

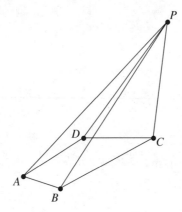

FIGURE 3

SECTION 15
Three Pretty Theorems in Geometry

The results in this essay were very kindly sent to me by Professor Juan-Bosco Romero Márquez of the University of Valladolid in Spain. It is a pleasure to congratulate Professor Márquez on his delightful discoveries and to thank him for permission to present them in my own words.

1. A Concurrency and a Collinearity

Let AD be an altitude in $\triangle ABC$, and let perpendiculars from D meet the other two sides in E and F (Figure 1). Also let parallels from D to the other sides meet them in G and H. Then

(a) the lines EF and GH are concurrent at a point A^* on BC extended, and

(b) letting B^* and C^* denote the corresponding points on the other sides of the triangle, the three points A^*, B^*, and C^* are collinear.

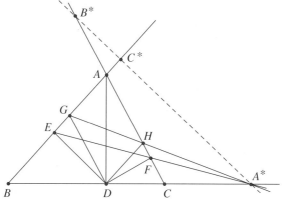

FIGURE 1

125

To establish part (a), it is enough to show that *EF* and *GH* divide *BC* externally in the same ratio. This is nicely accomplished by the theorem of Menelaus, a further application of which also yields part (b). Let us begin, then, by recalling the theorem of Menelaus.

The Theorem of Menelaus. *The points D, E, and F, on the sides AB, BC, and CA, respectively, of △ABC (Figure 2) are collinear if and only if the ratios of the directed segments*

$$\frac{AD}{DB} \cdot \frac{BE}{EC} \cdot \frac{CF}{FA} = -1.$$

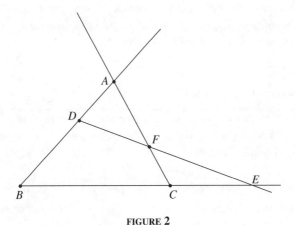

FIGURE 2

A proof of the theorem of Menelaus, and also of Ceva's theorem, which we will use later in this essay, can be found in my *Episodes from Nineteenth and Twentieth Century Euclidean Geometry*, The Anneli Lax New Mathematical Library Series, Vol. 37, 1995.

We note that if a point divides a side externally, the ratio of the parts is negative. Therefore, since a straight line must divide externally either one side or all three sides of a triangle, collinearity requires the product of the three ratios to be negative.

(a) Let *EF* meet *BC* externally at *X*, and let the lengths of the segments be as marked in Figure 3. Then $g = XC$ is negative, and since *EFX* is straight, the theorem of Menelaus yields

$$\frac{d}{e} \cdot \frac{f}{g} \cdot \frac{h}{i} = -1.$$

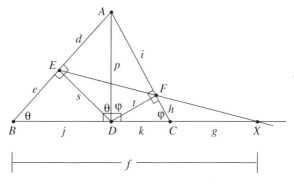

FIGURE 3

Since the altitude to the hypotenuse of a right triangle partitions the triangle into two smaller triangles each of which is similar to the original triangle, we have pairs of equal angles φ and θ as marked in Figure 3. Then

$$\tan\theta = \frac{d}{s} = \frac{s}{e} = \frac{p}{j}, \quad \text{from which} \quad \frac{d}{e} = \frac{d}{s}\cdot\frac{s}{e} = \left(\frac{p}{j}\right)^2,$$

and

$$\cot\varphi = \frac{h}{t} = \frac{t}{i} = \frac{k}{p}, \quad \text{from which} \quad \frac{h}{i} = \frac{h}{t}\cdot\frac{t}{i} = \left(\frac{k}{p}\right)^2.$$

Since

$$\frac{d}{e}\cdot\frac{f}{g}\cdot\frac{h}{i} = -1, \quad \text{then} \quad \left(\frac{p}{j}\right)^2\cdot\frac{f}{g}\cdot\left(\frac{k}{p}\right)^2 = -1,$$

and X divides BC externally in the ratio

$$\frac{f}{g} = -\left(\frac{j}{k}\right)^2.$$

In the case of GH (Figure 4), the parallel lines give the proportions

$$\frac{m}{n} = \frac{k}{j} = \frac{s}{t}.$$

By the theorem of Menelaus we have

$$\frac{m}{n}\cdot\frac{u}{v}\cdot\frac{s}{t} = -1,$$

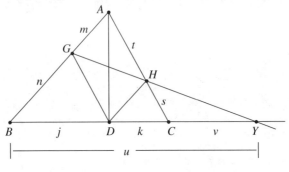

FIGURE 4

that is,

$$\frac{k}{j} \cdot \frac{u}{v} \cdot \frac{k}{j} = -1,$$

and

$$\frac{u}{v} = -\left(\frac{j}{k}\right)^2 \quad \left(= \frac{f}{g} \text{ in Figure 3}\right),$$

as desired.

(b) Now, in addition to the theorem of Menelaus, in this section we will also make use of the theorem of Ceva, both of which concern directed segments as the sides of a triangle are considered in cyclic order. Let us begin, then, by defining a cevian of a triangle and recalling Ceva's theorem.

Definition. A cevian of a triangle is a segment that joins a vertex to an interior or an exterior point of the opposite side.

Ceva's Theorem. *Cevians AD, BE, and CF of △ABC are concurrent if and only if the product of the directed ratios*

$$\frac{AF}{FB} \cdot \frac{BD}{DC} \cdot \frac{CE}{EA} = 1. \quad \text{(Figure 5)}$$

Again we omit the proof, which is given in most texts on elementary geometry. In contrast to the theorem of Menelaus, in Ceva's theorem concurrency implies an even number of external divisions of the sides, namely zero or two, making the product of the ratios positive.

Recall that we found in part (a) (Figure 3) that A^* divides BC externally in the ratio

Three Pretty Theorems in Geometry

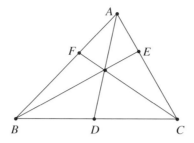

FIGURE 5

$$\frac{f}{g} = \frac{BA^*}{A^*C} = -\left(\frac{j}{k}\right)^2 = -\left(\frac{BD}{DC}\right)^2.$$

Observe that this ratio is associated with the directed side BC, not CB, and that the first vertex B occurs in both numerators BA^* and BD. Similarly, for the other sides and the altitudes BE and CF (Figure 6), we have that B^* divides CA in the ratio

$$\frac{CB^*}{B^*A} = -\left(\frac{CE}{EA}\right)^2 = -\left(\frac{q}{r}\right)^2$$

and C^* divides AB in the ratio

$$\frac{AC^*}{C^*B} = -\left(\frac{AF}{FB}\right)^2 = -\left(\frac{u}{v}\right)^2.$$

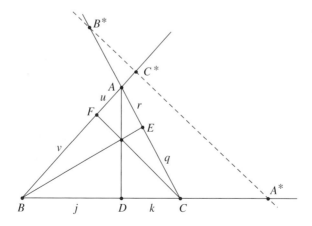

FIGURE 6

But the three altitudes are concurrent at the orthocenter. Hence by Ceva's theorem, we have

$$\frac{u}{v} \cdot \frac{j}{k} \cdot \frac{q}{r} = 1,$$

and so A^*, B^*, C^* divide the sides of $\triangle ABC$ in ratios whose product is

$$\left[-\left(\frac{u}{v}\right)^2\right] \cdot \left[-\left(\frac{j}{k}\right)^2\right] \cdot \left[-\left(\frac{q}{r}\right)^2\right] = -\left(\frac{u}{v} \cdot \frac{j}{k} \cdot \frac{q}{r}\right)^2 = -1,$$

establishing their collinearity by the theorem of Menelaus.

2. Cevian Conjugates

Let AD, BE, and CF be cevians through an arbitrary point P in the plane of $\triangle ABC$ (Figure 7). Through D and E draw lines parallel to cevian CF to meet side AB in M and L; similarly through E and F draw lines parallel to cevian AD to give R and Q on side BC, and through F and D draw lines parallel to cevian BE to give T and S on AC.

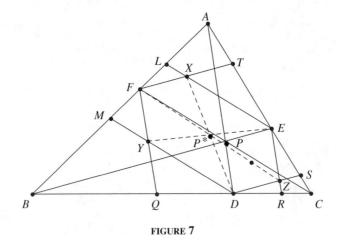

FIGURE 7

Let the intersection of FT and EL be X; similarly, let DM and FQ meet at Y, and DS and ER meet at Z. Then

(a) the segments XD, YE, and ZF are concurrent at a point P^*,
(b) P^* is the midpoint of each of these segments, from which it follows that

(c) a halfturn about P^* takes $\triangle XYZ$ to $\triangle DEF$, implying that the triangles are congruent.

P^* is called the cevian conjugate of P with respect to $\triangle ABC$.
The constructed parallel lines make quadrilaterals $FYDP$ and $FPEX$ parallelograms (Figure 8). Thus the sides YD and XE, being opposite the common side FP, are equal and parallel, implying $YDEX$ is also a parallelogram.

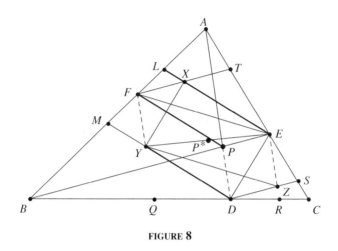

FIGURE 8

Similarly, $FYDP$ and $PDZE$ are parallelograms on opposite sides of PD, making FY and EZ equal and parallel and implying $FYZE$ is a parallelogram.

But parallelograms $YDEX$ and $FYZE$ have a common diagonal YE, and so the midpoint P^* of YE is also the midpoint of the other diagonals XD and FZ, and the theorem follows.

3. A Cevian Property

Let AD be an arbitrary cevian in $\triangle ABC$ (Figure 9). Lines from D, parallel to the other sides, meet them in E and F, and lines through E and F parallel to the cevian meet BC in G and H. Then

$$EG + FH = AD.$$

Clearly $EDFA$ is a parallelogram. Hence diagonal EF is bisected by diagonal AD, and since AD is parallel to EG and FH, it follows that AD is the

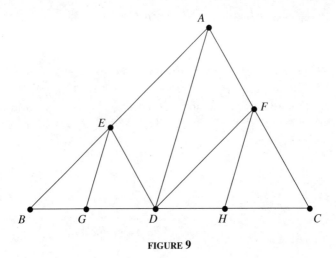

FIGURE 9

midline of the strip of the plane which is bounded by the parallels EG and FH. Thus AD bisects every transversal across the strip, in particular GH, and we have the little bonus that

$$GD = DH.$$

If FJ is drawn parallel to DH to meet AD at J (Figure 10), it is easy to see that triangles EGD and AJF are congruent (SAS):

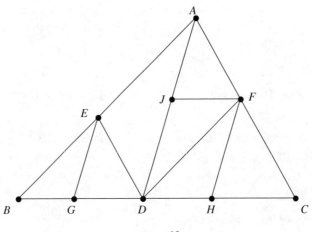

FIGURE 10

JDHF is a parallelogram, giving $JF = DH = GD$; in parallelogram *EDFA*, $AF = ED$; and the angles at D and F are equal since their respective sides are parallel.

Hence $EG = AJ$, and since $FH = JD$ in parallelogram *JDHF*, then

$$EG + FH = AJ + JD = AD.$$

SECTION 16
Two Gems from Euclidean Geometry*

1. The Butterfly Problem

The following is a famous old problem.

The Butterfly Problem. *Through the midpoint M of chord AB in a circle, arbitrary chords CD and EF are drawn (Figure 1). Prove that CF and ED cut off equal segments XM and MY along AB.*

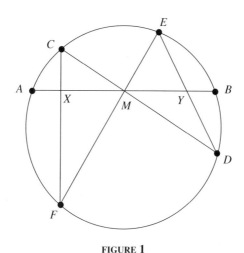

FIGURE 1

There are at least a couple of dozen solutions to this intriguing problem. First let us consider one of the neat standard solutions before going on to the sensational solution of my colleague Professor Hiroshi Haruki.

---
*This is a reworked version of material from my "Mathematical Gems" columns in the *Two-Year College Mathematics Journal*, Vol. 14, January and March, 1983.

Solution 1

Let the center of the circle be O, and let F be reflected in the diameter MO to F' on the circle (Figure 2). Then FF' is perpendicular to MO, $FM = F'M$, and the base angles 1 and 2 in isosceles triangle FMF' are equal.

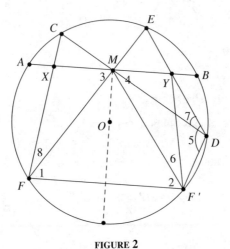

FIGURE 2

Since M is the midpoint of AB, diameter MO is also perpendicular to AB, making AB and FF' parallel. Thus alternate angles 1 and 3 are equal, as are angles 2 and 4. Since angles 1 and 2 are equal, then all four of these angles are equal. In triangles FXM and $F'YM$, then, we already have a pair of equal angles and a pair of equal sides, and one more appropriate fact would make them congruent, giving the desired $XM = MY$. It is easy to prove that angles 6 and 8 are equal as follows.

Let $\angle F'DE$ be labelled 5. Then, in cyclic quadrilateral $FF'DE$, opposite angles 1 and 5 are supplementary. Since angles 1 and 4 are equal, it follows that angles 4 and 5 are supplementary, making $MF'DY$ a cyclic quadrilateral. In the circumcircle around $MF'DY$, the chord MY subtends equal angles 6 and 7 at F' and D on the circumference. But angles 7 and 8 are equal angles subtended by chord CE in the given circle, and we have angle 6 = angle 8, completing the proof.

Haruki's Solution

The desired result is an easy corollary of the following lemma, which we shall prove shortly.

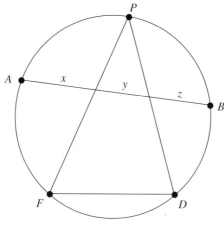

FIGURE 3

Haruki's Lemma. *Suppose AB and FD are nonintersecting chords in a circle and that P is a variable point on the arc AB remote from F and D (Figure 3). Then, for each position of P, the lines PF and PD cut AB into three segments of lengths x, y, z, which vary with P, but for which the value of xz/y remains constant.*

In the figure of the butterfly, then, let the lengths of the segments into which AB is divided by CF, M, and ED, respectively be w, x, y, and z (Figure 4). By Haruki's lemma, we have, for the point P at C and E, that

$$\frac{w(y+z)}{x} = \text{the constant}$$
$$= \frac{(w+x)z}{y}.$$

Since M is the midpoint of AB, we have $(w+x) = (y+z)$. Cancelling these equal factors, we get

$$\frac{w}{x} = \frac{z}{y}.$$

Adding 1 to each side gives

$$\frac{w+x}{x} = \frac{y+z}{y},$$

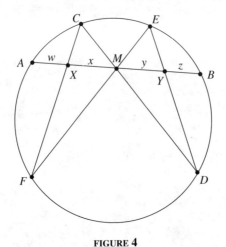

FIGURE 4

which again allows a similar cancellation to yield just

$$\frac{1}{x} = \frac{1}{y}$$

and the desired $x = y$.

Finally let us look at Professor Haruki's ingenious proof of his lemma. The proof is based on the well-known elementary theorem

equal products are determined by the parts of intersecting chords in a circle: $ab = cd$ (Figure 5).

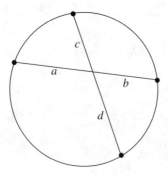

FIGURE 5

Let *PF* and *PD* cross *AB* at *Q* and *R*, respectively, and let the circumcircle of triangle *PQD* cross *AB* extended at *E* (Figure 6). Now, the angle $\theta = \angle FPD = \angle QPD$ remains constant as *P* varies on its arc, and the chord *QD* in the second circle subtends the same angle at *P* and *E*. Hence, for all positions of *P*, $\angle AED$ is always the same angle θ, implying that the circumcircles of all the triangles *PQR* cross *AB* extended at the same point *E*. Thus, for all *P*, the length of *BE* is some constant *k*.

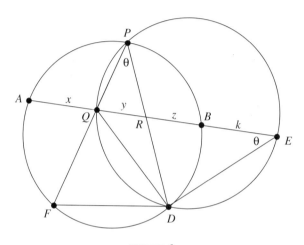

FIGURE 6

Now, the intersecting chords *AB* and *PD* in the given circle give the equal products

$$AR \cdot RB = PR \cdot RD,$$

and similarly in the second circle we have

$$PR \cdot RD = QR \cdot RE.$$

Hence

$$AR \cdot RB = QR \cdot RE,$$

that is,

$$(x+y)z = y(z+k),$$

giving

$$xz + yz = yz + yk,$$
$$xz = yk,$$

and the desired

$$\frac{xz}{y} = k.$$

2. The Remarkable Point Q

Suppose P is a point on a straight line m that does not intersect a given circle C. Obviously, the length of the tangent to C varies from point to point as P runs along m. Would you believe that

there exists a point Q such that the length of the tangent from P to C is always just the length of PQ? (Figure 7)

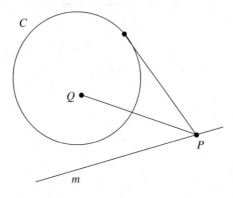

FIGURE 7

The perpendicular OM to m from the center O of the circle is clearly an axis of symmetry of the figure (Figure 8). Since P ranges on both sides of M, this marvellous point Q, if it exists, would be obliged to lie on the axis of symmetry. Thus there is no difficulty in locating Q if it exists; simply draw the tangent PT from any point P on m and mark Q on OM so that $PT = PQ$. Taking P at the point M (Figure 8), it remains only to prove that the length of the tangent RS from any other point R on m is equal to RQ, i.e., $t = v$.

Let the lengths of the segments be labelled as in Figure 8. Applying the theorem of Pythagoras to four right triangles, we obtain

Two Gems from Euclidean Geometry 141

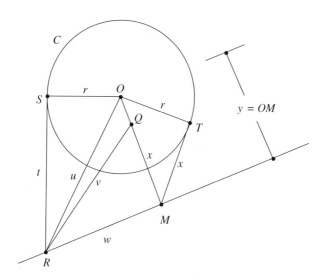

FIGURE 8

$$\text{from } \triangle MTO, \quad x^2 + r^2 = y^2, \tag{1}$$
$$\text{from } \triangle ROM, \quad y^2 + w^2 = u^2, \tag{2}$$
$$\text{from } \triangle RSO, \quad t^2 + r^2 = u^2, \tag{3}$$
$$\text{from } \triangle RMQ, \quad w^2 + x^2 = v^2. \tag{4}$$

Thus, from (3),

$$t^2 = u^2 - r^2 = (y^2 + w^2) - (y^2 - x^2) \quad \text{(from (2) and (1))}$$
$$= w^2 + x^2 = v^2 \quad \text{(from (4))},$$

giving the desired

$$t = v.$$

SECTION 17
The Thue–Morse–Hedlund Sequence

1. The Thue–Morse–Hedlund sequence is an infinite sequence of 0's and 1's $\{a_0, a_1, a_2, \ldots\}$ in which $a_n = 0$ if the number of 1's in the binary representation of n is even and $a_n = 1$ if the number of 1's is odd. From the table we see that the first 16 terms are

0 1 1 0 1 0 0 1 1 0 0 1 0 1 1 0 1 0 0 1 0 1 1 0 0 1 1 0 1 0 0 1.

n:	0	1	2	3	4	5	6	7
Base 2:	0	1	10	11	100	101	110	111
a_n:	0	1	1	0	1	0	0	1

n:	8	9	10	11	12	13	14	15
Base 2:	1000	1001	1010	1011	1100	1101	1110	1111
a_n:	1	0	0	1	0	1	1	0

Now, for $k \geq 1$, the number of $(k+1)$-digit binary integers is 2^k: the first digit must be 1; it is only in the last k places that there is a choice of 0 or 1. Thus each such integer $1ab \ldots c$ is equal to $2^k + r$, where $r = ab \ldots c$ is one of the 2^k smaller binary integers $\{0, 1, 10, 11, \ldots\}$ up to $2^k - 1$. Clearly $2^k + r$ has one more 1 in its binary representation than r has, and so $a_{2^k+r} = 1$ or 0 as $a_r = 0$ or 1. Thus the values of a_n for n equal to the $(k+1)$-digit binary integers, $a_{2^k}, a_{2^k+1}, \ldots, a_{2^k+(2^k-1)}\}$, are respectively the opposites of all the terms ahead of them in the sequence, $\{a_0, a_1, a_2, \ldots, a_{2^k-1}\}$. Therefore the sequence may be constructed from $a_0 = 0$ by repeatedly doubling the number of terms by adding a block consisting of the reversals of all the preceding terms:

0 1 : 1 0 : 1 0 0 1 : 1 0 0 1 0 1 1 0 : 1 0 0 1 0 1 1 0 0 1 1 0 1 0 0 1 ….

One of the first things we should like to know about the sequence is whether it repeats beyond some point. We shall prove in section 4, however, that the sequence is nonperiodic; in fact, the infinite binary decimal

143

determined by the sequence, 0.01101001..., has been shown to be a transcendental number!

Before proceeding, let us note two basic properties of the sequence. If the binary representation of n is $ab\ldots c$, then $2n$ is $ab\ldots c0$, containing the same number of 1's as n, and $2n+1$ is $ab\ldots c1$, containing one more 1 than n. Hence for all $n = 0, 1, 2, \ldots$,

$$(i) \quad a_{2n} = a_n \quad \text{and} \quad (ii) \quad a_{2n} \neq a_{2n+1}.$$

Note that (ii) does not imply a_{2n-1} and a_{2n} are different.

2. Because the sequence is constructed by reversing the terms in blocks from the beginning, among the first 2^{k+1} terms of the sequence there are 2^k terms equal to 0 and 2^k terms equal to 1. Now, suppose the 2^{k+1} integers $\{0, 1, 2, 3, 4, \ldots, 2^{k+1} - 1\}$ are divided into two subsets X_k and Y_k so that X_k contains the 2^k integers n for which $a_n = 0$ and Y_k contains the 2^k integers n for which $a_n = 1$. For example, for $k = 3$ we have

$$X_3 = \{0, 3, 5, 6, 9, 10, 12, 15\}, \quad Y_3 = \{1, 2, 4, 7, 8, 11, 13, 14\}.$$

In this case, then

$$0^0 + 3^0 + 5^0 + 6^0 + 9^0 + 10^0 + 12^0 + 15^0$$
$$= 1^0 + 2^0 + 4^0 + 7^0 + 8^0 + 11^0 + 13^0 + 14^0 \quad (=8),$$
$$0^1 + 3^1 + 5^1 + 6^1 + 9^1 + 10^1 + 12^1 + 15^1$$
$$= 1^1 + 2^1 + 4^1 + 7^1 + 8^1 + 11^1 + 13^1 + 14^1 \quad (=60),$$
$$0^2 + 3^2 + 5^2 + 6^2 + 9^2 + 10^2 + 12^2 + 15^2$$
$$= 1^2 + 2^2 + 4^2 + 7^2 + 8^2 + 11^2 + 13^2 + 14^2 \quad (=620),$$
$$0^3 + 3^3 + 5^3 + 6^3 + 9^3 + 10^3 + 12^3 + 15^3$$
$$= 1^3 + 2^3 + 4^3 + 7^3 + 8^3 + 11^3 + 13^3 + 14^3 \quad (=7200).$$

It is surprisingly easy to prove by induction that, if $X_k = \{x_0, x_1, \ldots\}$, and $Y_k = \{y_0, y_1, \ldots\}$, then, for all exponents $t = 0, 1, 2, \ldots, k$,

$$x_0^t + x_1^t + \cdots = y_0^t + y_1^t + \cdots.$$

As if this weren't enough, we shall see it is no harder to prove the amazing theorem that, for any polynomial $f(x)$ of degree $\leq k$,

$$\sum_{n \in X_k} f(n) = \sum_{n \in Y_k} f(n).$$

The Thue–Morse–Hedlund Sequence

Proof: For $k = 0$, $f(x)$ is a constant, and the claim holds trivially since X_k and Y_k have the same number of members (in this case, just one). Suppose, then, that the claim holds for some $k \geq 0$ and that $f(x)$ is a polynomial of degree not exceeding $k + 1$. Let $g(x)$ denote the difference

$$g(x) = f(x + 2^{k+1}) - f(x).$$

Then $g(x)$ is a polynomial of degree $\leq k$ since any terms in x^{k+1} cancel, and the induction hypothesis gives

$$\sum_{n \in X_k} g(n) = \sum_{n \in Y_k} g(n),$$

that is,

$$\sum_{n \in X_k} f(n + 2^{k+1}) - \sum_{n \in X_k} f(n) = \sum_{n \in Y_k} f(n + 2^{k+1}) - \sum_{n \in Y_k} f(n).$$

Transposing, gives

$$\sum_{n \in X_k} f(n + 2^{k+1}) + \sum_{n \in Y_k} f(n) = \sum_{n \in Y_k} f(n + 2^{k+1}) + \sum_{n \in X_k} f(n),$$

and it remains only to observe that we already have precisely the result we want, namely

$$\sum_{n \in Y_{k+1}} f(n) = \sum_{n \in X_{k+1}} f(n).$$

On the left side, f is to be evaluated at each member n of Y_k and also at the numbers $n + 2^{k+1}$, where $n \in X_k$; but, altogether, these are just the members of Y_{k+1}. Recall that Y_k consists of the nonnegative integers i less than 2^{k+1} for which $a_i = 1$, and that Y_{k+1} comprises all such integers less than 2^{k+2}. Thus $Y_k \subset Y_{k+1}$.

Now, $n + 2^{k+1}$, being at least 2^{k+1}, is too big for Y_k, but, with n belonging to X_k, we have $n < 2^{k+1}$, implying $n + 2^{k+1} < 2^{k+2}$; moreover, with $n \in X_k$, then $a_n = 0$, in which case $a_{n+2^{k+1}} = 1$, making $n + 2^{k+1}$ a full-fledged member of Y_{k+1}. In order of magnitude, the members of Y_{k+1} are

$$\{\ldots \text{the } 2^k \text{ members of } Y_k \ldots, \ldots \text{the } 2^k \text{ numbers } n + 2^{k+1},$$
$$\text{where } n \in X_k, \ldots\}.$$

Similarly, on the right side f is evaluated at the members of X_{k+1}, which is similarly seen to be

$$\{\ldots \text{the } 2^k \text{ members of } X_k \ldots, \ldots \text{the } 2^k \text{ numbers } n + 2^{k+1},$$
$$\text{where } n \in Y_k, \ldots\}.$$

The theorem follows by induction.

In particular, taking $f(x) = x^t$ in turn for $t = 0, 1, \ldots, k$ we obtain the set of $k+1$ equations in question. This property of the sequence was observed by M. E. Prouhet as early as 1851.

3. Next, let X_k and Y_k be extended to the infinite sets X and Y which respectively contain all the nonnegative integers n for which $a_n = 0$ and $a_n = 1$. Then we have another remarkable property:

> the number of ways of expressing a nonnegative integer n as a sum of two distinct members of X is the same as the number of ways of expressing it as a sum of two distinct members of Y;

> moreover, the partition (X, Y) is the only way of dividing the integers $0, 1, 2, \ldots$ into two disjoint subsets that have this property.

From X_4 and Y_4 we can check a few small values:

$$X_4 = \{0, 3, 5, 6, 9, 10, 12, 15, 17, 18, 20, 23, 24, 27, 29, 30\},$$
$$Y_4 = \{1, 2, 4, 7, 8, 11, 13, 14, 16, 19, 21, 22, 25, 26, 28, 31\}.$$

For example, there are two ways of expressing 25 as a sum of distinct members of X_4, namely $5 + 20$ and $10 + 15$, and two ways as a sum of distinct members of Y_4, $4 + 21$ and $11 + 14$. We observe that there may be no way of expressing certain integers as such sums; for example, 4 or 7.

The proof of this theorem, which was established by the Canadian mathematicians Leo Moser and Joe Lambek in 1959, is a straightforward elementary application of generating functions.

Let us first show that if (A, B) is a partition of $\{0, 1, 2, \ldots\}$ which has the property in question, then $A = X$ and $B = Y$. Suppose, then, that (A, B) is such a partition and, for definiteness, that $0 \in A$:

$$A = \{0, a_1, a_2, \ldots\}, \quad B = \{b_0, b_1, b_2, \ldots\}.$$

Now let us define the generating functions

$$A(x) = x^0 + x^{a_1} + x^{a_2} + \cdots = 1 + x^{a_1} + x^{a_2} + \cdots,$$

and

$$B(x) = x^{b_0} + x^{b_1} + x^{b_2} + \cdots.$$

Since A and B contain all the nonnegative integers between them, we have
$$A(x) + B(x) = 1 + x + x^2 + x^3 + \cdots = \frac{1}{1-x}.$$
Next, consider the product
$$A^2(x) = (1 + x^{a_1} + x^{a_2} + \cdots)(1 + x^{a_1} + x^{a_2} + \cdots)$$
$$= \cdots + c_n x^n + \cdots.$$
The coefficient c_n here is the number of ways that the exponent n can be expressed as a sum $a_s + a_t$ of two members of A. Clearly, when n is even, this might include the case when a_s and a_t are the same. Since we want a_s and a_t to be distinct, we need to eliminate each of the terms x^{2a_s}. Now $x^{2a_s} = (x^2)^{a_s}$ and these unwanted terms are simply those in
$$A(x^2) = 1 + x^{2a_1} + x^{2a_2} + \cdots.$$
Hence the number of ways of obtaining n as a sum of two distinct members of A is the coefficient of x^n in the series
$$A^2(x) - A(x^2).$$

Similarly, the coefficient of x^n in $B^2(x) - B(x^2)$ is the number of ways of expressing n as a sum of two distinct members of B. Since these coefficients are given to be the same for all n, the two series must be identical:
$$A^2(x) - A(x^2) = B^2(x) - B(x^2).$$
Hence
$$A^2(x) - B^2(x) = A(x^2) - B(x^2),$$
which factors to give
$$[A(x) + B(x)][A(x) - B(x)] = A(x^2) - B(x^2).$$
Recalling that $A(x) + B(x) = \frac{1}{1-x}$, we get the important relation
$$A(x) - B(x) = (1-x)[A(x^2) - B(x^2)].$$
Replacing x in turn by $x^2, x^4, \ldots$, this yields
$$A(x^2) - B(x^2) = (1-x^2)[A(x^4) - B(x^4)],$$
$$A(x^4) - B(x^4) = (1-x^4)[A(x^8) - B(x^8)],$$
$$\cdots\cdots\cdots.$$

Thus

$$A(x) - B(x) = (1-x)(1-x^2)\cdots(1-x^{2^i})[A(x^{2^{i+1}}) - B(x^{2^{i+1}})],$$

where the string of binomial factors on the right side can be made as long as we please.

Now, except for the $x^0 = 1$ which begins the series $A(x^{2^{i+1}})$, every power of x in $[A(x^{2^{i+1}}) - B(x^{2^{i+1}})]$ is of degree at least 2^{i+1} and plays no part in the construction of any term x^n of lesser degree. Since 2^{i+1} becomes arbitrarily large with increasing i, every term x^n is of degree less than 2^{i+1} for some i. Thus every term of the series $A(x) - B(x)$ generated by multiplying out the infinite product $(1-x)(1-x^2)\cdots(1-x^{2^i})\ldots$, and we have

$$A(x) - B(x) = (1-x)(1-x^2)\cdots(1-x^{2^i})\ldots.$$

In multiplying out the right side, each term carries a coefficient of ± 1, the positive terms belonging to $A(x)$ and the negative ones to $B(x)$. Now, each exponent is a power of 2, and a term $(-x^{2^r})(-x^{2^s})\cdots(-x^{2^t}) = +x^n$ if and only if it contains an even number of factors, that is, if and only if the exponent $n = 2^r + 2^s + \cdots + 2^t$ is the sum of an even number of distinct powers of 2. Each x^n in $A(x)$, then, carries an exponent n which has an even number of 1's in its binary representation, making n a member of X. Similarly, each $-x^n$ in $B(x)$ implies $n = 2^c + 2^d + \cdots + 2^e$, containing an odd number of distinct powers of 2, and $n \in Y$. The conclusion follows.

Thus we have established that if such sequences A and B exist, it must be that $A = X$ and $B = Y$. To complete the proof of the theorem we need also to show the converse, namely if $A = X$ and $B = Y$, then the sequences A and B do enjoy the property in question.

Therefore, suppose $A = X$ and $B = Y$. Expanding the infinite product

$$(1-x)(1-x^2)(1-x^4)\cdots(1-x^{2^n})\ldots,$$

the exponents of the positive terms constitute the sequence X and the exponents of the negative terms constitute Y. Recalling that the generating functions $A(x)$ and $B(x)$ draw their exponents from the sets A and B, respectively, we obtain

$$(1-x)(1-x^2)(1-x^4)\cdots(1-x^{2^n})\cdots = A(x) - B(x). \tag{1}$$

Also, since X and Y contain all the nonnegative integers, we again have

$$A(x) + B(x) = 1 + x + x^2 + x^3 + \cdots = \frac{1}{1-x}.$$

The Thue–Morse–Hedlund Sequence

Therefore

$$[A(x) - B(x)][A(x) + B(x)] = (1 - x^2)(1 - x^4) \cdots (1 - x^{2^n}) \ldots,$$
$$= A(x^2) - B(x^2) \quad \text{from (1) above.}$$

Hence

$$A^2(x) - B^2(x) = A(x^2) - B(x^2),$$

implying

$$A^2(x) - A(x^2) = B^2(x) - B(x^2),$$

which we recognize as the condition for A and B to possess the property in question, and the proof is complete.

Lambek and Moser also gave the following partition (C, D) which has the same property with regard to multiplication instead of addition:

$$C = \{1, 6, 8, 10, 12, 14, 15, 18, 20, 21, 22, 26, 27, 28, \ldots\}$$
$$D = \{2, 3, 4, 5, 7, 9, 11, 13, 16, 17, 19, 23, 24, 25, 29, 30, \ldots\}.$$

The rule of formation is the following:

if $n = p_1^{a_1} p_2^{a_2} \ldots p_t^{a_t}$ is the prime decomposition of n, then, expressing all the exponents a_i in binary notation, n goes into C if the total number of 1's in all the a's is even and into D if it is odd.

4. Finally, let us show that the Thue–Morse–Hedlund sequence is nonperiodic.

Suppose, to the contrary, that, beyond some term a_v, the sequence repeats with a smallest period equal to k. If k is even, say $k = 2t$, then for $n > v$, we have, recalling that $a_{2n} = a_n$ for all n, that

$$a_n = a_{2n} = a_{2n+k} = a_{2n+2t} = a_{2(n+t)} = a_{n+t},$$

giving the contradiction that the sequence is also periodic with the smaller period t.

If k is odd, consider a string of k consecutive terms

$$a_m, a_{m+1}, a_{m+2}, \ldots, a_{m+k-1},$$

selected far enough along the sequence so that a_{m-1} is also in the periodic part of the sequence. Since it contains an odd number of terms, there can't be the same number of 0's as 1's in the string. For definiteness, suppose there is at least one more 0 than 1. Now, since k is the period, the next k terms in the sequence merely duplicate this string and hence they abut to give a string of length $2k$,

$$a_m, a_{m+1}, \ldots, a_{m+k-1}, a_{m+k}, \ldots, a_{m+2k-1},$$

which contains at least two more 0's than 1's.

Recalling that $a_{2n} \neq a_{2n+1}$ for all n, it follows that each pair (a_{2n}, a_{2n+1}) contains a 0 and a 1. Hence if m is even, then the string consists simply of k abutting pairs of this type, implying the contradiction that the string must contain the same number of 0's as 1's.

However, if m is odd, the first and last terms in the string are not contained in such pairs:

$$a_m, (a_{m+1}, a_{m+2}), \ldots, (a_{m+2k-3}, a_{m+2k-2}), a_{m+2k-1}.$$

But, for m odd, we have $a_{m-1} \neq a_m$ (again using $a_{2n} \neq a_{2n+1}$), and since $a_{m-1} = a_{m+2k-1}$ because k is the period, we have $a_m \neq a_{m+2k-1}$. Thus the first and last terms are different, again implying the contradiction that the string must contain the same number of 0's as 1's. Thus it follows by contradiction that the sequence cannot be periodic.

Although the sequence was known earlier, it is called the Thue–Morse–Hedlund sequence because of the special attention it received from the Norwegian mathematician Axel Thue in the first decade of the twentieth century and in a paper in the 1940s by M. Morse and G. A. Hedlund.

This essay is based on material from two outstanding books, each of which contains a wealth of fascinating topics:

(i) *Mathematical Miniatures*, by Svetoslav Savchev and Titu Andreescu (The Anneli Lax New Mathematical Library Series, MAA, 2003)
(ii) *Lure of the Integers*, by Joe Roberts (Spectrum Series, MAA, 1992).

SECTION 18
Two Miscellaneous Problems

These two problems came my way from Riko Winterle of Waterloo, who was in a course or two of mine some thirty years ago and who, though not a professional mathematician, has always been fascinated with the challenges and rewards of mathematical activity. Riko formulated and solved these problems on his own, without knowing that the second problem is equivalent to a result of long standing in elementary analysis.

1. Recall that the "medial" triangle of $\triangle ABC$ is determined by the midpoints A', B', C' of its sides (Figure 1). It is well known that the medial triangle partitions $\triangle ABC$ into four triangles of equal area (each side of $\triangle A'B'C'$ is a diagonal of a parallelogram, making the medial triangle congruent to each of the other three).

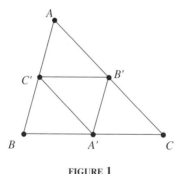

FIGURE 1

The converse is an interesting problem (Figure 2):

if an inscribed triangle *DEF*, having a vertex in the interior of each side of $\triangle ABC$, divides the triangle into four equal triangles, must the vertices D, E, F, be the midpoints of the sides?

151

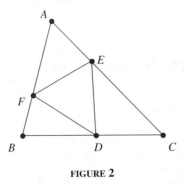

FIGURE 2

In other words, is the medial triangle the only inscribed triangle that partitions the triangle into four equal triangles?

Clearly D lies in BA' or in $A'C$ (not at B or C, of course). For definiteness, suppose D lies in $A'C$. Now, if any vertex of $\triangle DEF$ were to coincide with a midpoint of a side, it follows easily that $\triangle DEF$ would have to be the medial triangle: for example,

if D is at A' (Figure 3), then, unless E is at B', $\triangle CDE$ would be greater or smaller than triangle $CA'B'$, depending on which side of B' vertex E is on. Similarly, in order to have each triangle $(\frac{1}{4})\triangle ABC$, F would have to coincide with C', and DEF would be the medial triangle.

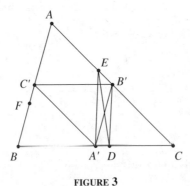

FIGURE 3

Moreover, observe that if D is in the interior of $A'C$, then E must be in the interior of $B'A$ (Figure 3):

Two Miscellaneous Problems

subtracting from the equal triangles CDE and $CA'B'$ their common part, $\triangle CDB'$, we get $\triangle DEB' = \triangle DA'B'$, implying $A'E$ and DB' are parallel, forcing DE and $A'B'$ to cross internally.

Thus, with D in the interior of $A'C$, E must lie in the interior of $B'A$ and similarly F in the interior of $C'B$ (Figure 4).

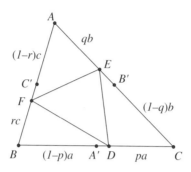

FIGURE 4

As usual, let the lengths of the sides of $\triangle ABC$ be denoted by a, b, c. Then, allowing for the possibility of D coinciding with A' we have $DC = pa$ for some p in $0 < p \le \frac{1}{2}$, and the complementary segment $BD = a - pa = (1-p)a$. Similarly, $EA = qb$ and $FB = rc$, for some q, r in $0 < q, r \le \frac{1}{2}$, with the complementary segments of the sides being $(1-q)b$ and $(1-r)c$ as in Figure 4.

Using the formula $\frac{1}{2}xy \sin Z$ for the area of a triangle, we have from

$$\triangle DCE = \tfrac{1}{4} \triangle ABC$$

that

$$\tfrac{1}{2} pa(1-q)b \sin C = \tfrac{1}{4}\left(\tfrac{1}{2} ab \sin C\right).$$

Dividing by $\frac{1}{2}ab \sin C$, this gives

$$p(1-q) = \tfrac{1}{4}. \qquad (1)$$

Similarly, at the other vertices we obtain

$$q(1-r) = \tfrac{1}{4} \qquad (2)$$

and

$$r(1 - p) = \tfrac{1}{4}. \qquad (3)$$

Solving (1) for q, we have $1 - q = 1/4p$, giving

$$q = 1 - \frac{1}{4p} = \frac{4p - 1}{4p}.$$

Substituting in (2), we get

$$\frac{4p - 1}{4p}(1 - r) = \frac{1}{4},$$

from which

$$(1 - r) = \frac{p}{4p - 1},$$

and

$$r = 1 - \frac{p}{4p - 1} = \frac{3p - 1}{4p - 1}.$$

Finally, substituting in (3) yields

$$\frac{3p - 1}{4p - 1}(1 - p) = \frac{1}{4},$$
$$(12p - 4)(1 - p) = 4p - 1,$$

which easily reduces to

$$3(4p^2 - 4p + 1) = 3(2p - 1)^2 = 0,$$

and we have

$$p = \tfrac{1}{2}.$$

Thus D is the midpoint of BC and, as we saw above, it follows that DEF is the medial triangle.

2. Our second problem concerns subsets S of points on the real line. If S is bounded and its cardinality is infinite, the Bolzano-Weierstrass theorem guarantees that S has a point of accumulation P. However, it does not guarantee that P belongs to S. In this problem we are asked to show that, bounded or not, uncountability is a guarantee of this:

show that an *uncountable* subset S of the real line has a point of accumulation that *belongs to S*.

Observe that there are only two kinds of points in S—isolated points and points of accumulation:

consider an interval I centered at a point x of S. If, by making I sufficiently small, all the other points of S can be squeezed out of the interval, then x is an isolated point of S; otherwise *every* interval about x contains a point of S besides x, and x is a point of accumulation of S.

Thus it suffices to show that an uncountable set S cannot consist entirely of isolated points. Accordingly, let us show that the isolated points cannot amount to more than a countable subset of S.

Recall that the cardinality of the union of a countable number of countable sets is still countable. Consequently, since there is only a countable number of unit intervals $[n, n+1]$ along the real line, if each were to intersect S in only a countable set, then S could not amount to an uncountable set. Hence some interval $[n, n+1]$ must intersect S in an uncountable subset T. If we can show that T contains only a countable number of isolated points of S, then, even if every unit interval were to do the same, the isolated points of S would only amount to a countable subset. Let us proceed, then, with the central issue of showing that T can never contain more than a countable number of isolated points.

Let $T(r)$ be the subset of isolated points x in T which have the property that an interval of radius r, centered at x, is small enough to isolate x from the other points of S. Now, since T is contained in some *unit* interval $[n, n+1]$, if $T(r)$ were to contain more than $\frac{1}{r}+1$ points, the pigeonhole principle would imply the contradiction that some two of its points would be closer together than r. Thus the number of points in $T(r)$ must be finite, and therefore countable.

Consider now the net determined by the subsets $T(\frac{1}{n})$, $n = 2, 3, 4, 5, \ldots$. Since a point must possess some "T-free" neighborhood in order to qualify as an isolated point of T, it is clear that by taking $\frac{1}{n}$ small enough, i.e., n large enough, no isolated point of T can avoid belonging to some subset $T(\frac{1}{n})$. Thus the isolated points of T are all present in the union of the subsets $T(\frac{1}{n})$. But there is only a countable number of subsets $T(\frac{1}{n})$, and each contains only a countable number of points. Thus the isolated points in T are indeed countable, and the argument is complete.

An Alternative Approach

We conclude with a neat alternative approach to the central issue of the problem that is due to my long-time colleague Frank Zorzitto.

If we could pair up each isolated point x of S with a different rational point, then the isolated points couldn't amount to more than the cardinality of the rational points, which is known to be countable. Consider the following scheme.

Centered at an isolated point x is some iterval I_x, of positive length, which contains no point of S besides x. Let the radius of I_x be $2a$. Since every interval on the line contains a rational point (in fact a countable infinity of them), let the associated rational point p_x be chosen from the *interior* of the half-interval with center x and radius a (Figure 5). If a rational point is chosen in this way for each isolated point, it is easy to see that no rational point p_x could be assigned more than once.

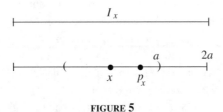

FIGURE 5

Let p_x and p_y be the rational points thus associated with two distinct isolated points x and y of S. Since the interval I_x contains no point of S besides x, in particular the point y, it follows that p_x is closer to x than one-half the distance between x and y. Similarly, p_y is likewise assigned so that it is closer to y than one-half the distance from x to y. Thus p_x and p_y lie inside different halves of the segment joining x and y, making them different points, and implying that no rational point is assigned more than once. The conclusion follows.

SECTION 19
A Surprising Property of Regular Polygons

It is a pleasure to thank John Bull of Great Brington, Northampton, UK, for directing my attention to the remarkable theorem which is the subject of this essay. Before considering the general n-gon, let us look at the interesting case of the pentagon.

1. The Pentagon

Let $ABCDE$ be a regular pentagon inscribed in a unit circle. Prove that the product of the lengths of the two sides and the diagonals from a vertex is 5, that is, in Figure 1, prove that

$$AB \cdot AC \cdot AD \cdot AE = 5.$$

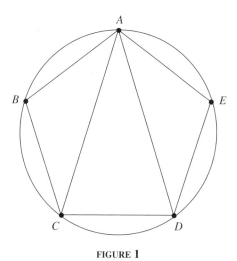

FIGURE 1

Because of the symmetry, we have $AB = AE$ and $AC = AD$. If we let $AB = x$ and $AC = y$ (Figure 2), then the product in question is

$$xyyx = (xy)^2.$$

157

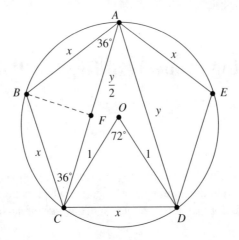

FIGURE 2

Hence we would like to show that

$$xy = \sqrt{5}.$$

Since each side of the pentagon subtends an angle of $(2\pi/5) = 72°$ at the center O, it subtends an angle of $36°$ at the circumference. Thus angles BAC and ACB are $36°$ in isosceles triangle ABC.

Applying the law of cosines to $\triangle OCD$ we get

$$x^2 = 2 - 2\cos 72° = 2 - 2(1 - 2\sin^2 36°) = 4\sin^2 36°.$$

In isosceles triangle ABC, whose base angles are $36°$, the altitude from B meets the base AC at its midpoint F. Thus

$$AF = x\cos 36° = \frac{y}{2},$$

giving

$$y = 2x\cos 36°.$$

The task of showing that $xy = \sqrt{5}$ thus reduces to proving

$$2x^2 \cos 36° = \sqrt{5},$$

that is,

$$8\sin^2 36° \cos 36° = \sqrt{5}.$$

It remains only to calculate these trig functions of $36°$.

A Surprising Property of Regular Polygons   159

For these values we turn to Euclid's method of inscribing a regular decagon in a circle. A corollary to this brilliant work is that

$$\cos 36° = \frac{1+\sqrt{5}}{4}.$$

This comes from a special property of the isosceles triangle which has a vertical angle of 36° (and hence base angles of 72°). Bisecting a base angle cuts from the original triangle another triangle that is similar to it. In Figure 3, triangles PQR and SQR are similar isosceles triangles. Clearly triangle PQS is yet another isosceles triangle (with $PS = QS$).

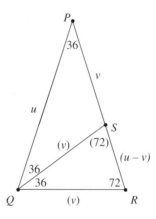

FIGURE 3

Letting

$$PQ = PR = u, \quad PS = v,$$

we have

$$QS = PS = v, \quad QR = QS = v, \quad \text{and} \quad SR = u - v.$$

From the proportional sides in triangles PQR and SQR, then,

$$\frac{PQ}{QR} = \frac{QR}{SR}, \quad \text{that is,} \quad \frac{u}{v} = \frac{v}{u-v}.$$

Hence

$$u(u-v) = v^2$$

and, dividing by v^2,

$$\frac{u}{v}\left(\frac{u}{v}-1\right)=1.$$

Now, triangle PQS is isosceles, and therefore the altitude from S meets base PQ at its midpoint. That is to say,

$$v\cos 36° = \frac{u}{2},$$

from which

$$\frac{u}{v} = 2\cos 36°.$$

Hence

$$\frac{u}{v}\left(\frac{u}{v}-1\right)=1$$

gives

$$2\cos 36°(2\cos 36° - 1) = 1,$$
$$4\cos^2 36° - 2\cos 36° - 1 = 0,$$

and

$$\cos 36° = \frac{2 \pm \sqrt{4+16}}{8} = \frac{1 \pm \sqrt{5}}{4}.$$

Since $\sqrt{5} > 1$ and $\cos 36°$ is positive, it follows that $\cos 36° = (1+\sqrt{5})/4$.

Hence

$$\cos^2 36° = \frac{6+2\sqrt{5}}{16} = \frac{3+\sqrt{5}}{8}$$

and

$$\sin^2 36° = 1 - \cos^2 36° = 1 - \frac{3+\sqrt{5}}{8} = \frac{5-\sqrt{5}}{8}.$$

Then the required product

$$xy = 8\sin^2 36° \cos 36°$$
$$= 8 \cdot \frac{5-\sqrt{5}}{8} \cdot \frac{1+\sqrt{5}}{4}$$
$$= \frac{5 - 5 + 5\sqrt{5} - \sqrt{5}}{4}$$
$$= \frac{4\sqrt{5}}{4} = \sqrt{5},$$

as desired.

A Surprising Property of Regular Polygons 161

Now let us turn to the general case.

2. The General Case

It is remarkable that this property holds in general:
If $A_0A_1 \cdots A_{n-1}$ is a regular n-gon inscribed in a unit circle, $n \geq 3$, then $A_0A_1 \cdot A_0A_2 \cdot A_0A_3 \cdots \cdots A_0A_{n-1} = n$.

Happily, this theorem is much easier to prove in general than for the pentagon.

The following standard factoring is easily verified by multiplying out the right side:

$$x^n - 1 = (x-1)(x^{n-1} + x^{n-2} + \cdots + x + 1).$$

Now, the roots of $x^n - 1 = (x-1)(x^{n-1} + x^{n-2} + \cdots + x + 1) = 0$ are the n nth roots of unity, namely

$$1, \omega, \omega^2, \ldots, \omega^{n-1}, \quad \text{where} \quad \omega = e^{2\pi i/n} = \cos\frac{2\pi}{n} + i\sin\frac{2\pi}{n}.$$

Clearly the factor $x - 1$ accounts for the root 1, and so the roots of

$$x^{n-1} + x^{n-2} + \cdots + x + 1 = 0 \quad \text{are} \quad \omega, \omega^2, \ldots, \omega^{n-1},$$

and yield the factoring

$$x^{n-1} + x^{n-2} + \cdots + x + 1 = (x - \omega)(x - \omega^2)\ldots(x - \omega^{n-1}).$$

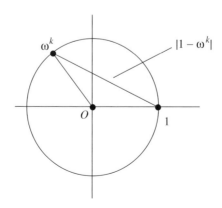

FIGURE 4

For $x = 1$, this gives
$$n = (1 - \omega)(1 - \omega^2) \cdots (1 - \omega^{n-1}),$$
and taking absolute values, we have
$$n = \left|1 - \omega\right|\left|1 - \omega^2\right| \cdots \left|1 - \omega^{n-1}\right|.$$
But this is already the result we want!

To see this, recall that, in the complex plane, the points $1, \omega, \omega^2, \ldots, \omega^{n-1}$ are the vertices of a regular n-gon inscribed in the unit circle and that the length of the chord joining the points 1 and ω^k is $|1 - \omega^k|$ (Figure 4).

SECTION **20**
Three Short Stories in Number Theory*

1. (From the journal $\pi\mu\varepsilon$, 1974, problem 334, page 30)

   If the following infinite series S is evaluated as a decimal, what is the digit in the 37th place?

   $$S = \frac{1}{9} + \frac{1}{99} + \frac{1}{999} + \cdots + \frac{1}{10^n - 1} + \cdots .$$

   Let S be evaluated by stacking the repeating decimals representing its terms one below the other with their decimal points in line.

   .111111111...
   .010101010...
   .001001001...
   ...

   The contributions to the kth decimal place consist only of 0's and 1's, and the 1's in column k occur in the rth row if and only if r is a *divisor* of k. Below the kth row in column k, only 0's occur. Since 37 is a prime number, its only divisors are 1 and 37, and so the only 1's in column 37 occur in rows 1 and 37. Thus it is tempting to jump to the conclusion that the required digit is 2. However, because sums are performed from right to left, we must consider the possibility of a carryover from the 38th place.

   Since 38 has four divisors, 1, 2, 19 and 38, there are four 1's in column 38. Of course, we must also be alert to the possibility of a carryover from the 39th place. It seems as if this is going to go on forever. However, since the 38th place holds only four 1's of its own, it would take a carryover of at least 6 from the 39th place to force the 38th place into a carryover. We shall show that the carryover from the 39th place doesn't even amount to 5, implying that there is no carryover from the 38th place and that the final digit in the 37th place is indeed 2.

---

*This is a reworked version of material from my "Mathematical Gems" column in the *Two-Year College Mathematics Journal*, Vol. 12, January, 1981.

This follows nicely by considering the situation to be worse than it really is. The 1's in column k occur only in the top k rows, and it would be impossible to build up a bigger carryover to the 38th place than that which would accrue in the event that every column k beyond the 38th were to contain nothing but 1's in its top k rows. Noting that the value of a 1 in column k is $\frac{1}{10^k}$, in such a case the value of all the columns beyond the 38th would be given by the infinite series

$$T = \frac{39}{10^{39}} + \frac{40}{10^{40}} + \frac{41}{10^{41}} + \cdots.$$

Now, for $k \geq 39$, it is clear that $k + 1 < 2k$. Hence, for $k \geq 39$, we have

$$\frac{k+1}{10^{k+1}} < \frac{2k}{10^{k+1}} = \frac{1}{5} \cdot \frac{k}{10^k}.$$

This tells us that each term in T is less than one-fifth of the previous term, and we have

$$T < \frac{39}{10^{39}} + \frac{1}{5}\left(\frac{39}{10^{39}}\right) + \frac{1}{5^2}\left(\frac{39}{10^{39}}\right) + \cdots$$

$$= \frac{39/10^{39}}{1 - \frac{1}{5}} = \frac{5}{4} \cdot \left(\frac{39}{10^{39}}\right) < \frac{5}{4} \cdot \left(\frac{40}{10^{39}}\right) = \frac{5}{10^{38}}.$$

Thus the maximum possible carryover T does not even amount to a 5 in the 38th place and the proof is complete.

2. Investigations into the question of how the prime numbers are distributed among the positive integers have uncovered many surprising facts. It is well known, for example, that there exist arbitrarily long stretches of consecutive positive integers which are all composite. The following remarkable property, however, seems not to be so widely known: There exists an infinity of prime numbers which occur oasis-like in the middle of a string of composite integers. In fact, such a string can be made arbitrarily long:

for each positive integer n, there is a prime number p such that, in the increasing sequence of positive integers, the n consecutive integers immediately to the left of p and the n consecutive integers immediately to the right of p are all composite.

The beautiful proof below, due to the great Polish mathematician Waclaw Sierpinski, is given in the highly recommended book *Matters Mathematical* by Herstein and Kaplansky (Harper and Row, 1974, pages 39–40).

The proof is based on Dirichlet's famous theorem

if a and b are relatively prime positive integers, the arithmetic progression $\{am + b\}$ contains an infinity of prime numbers.

Suppose $n > 1$ is specified. Let q be any prime that exceeds n by at least 2, i.e., $q - n \geq 2$, and consider the product "a" of the n positive integers on each side of it:

$$a = (q - n)\big[q - (n - 1)\big] \cdots (q - 1)(q + 1)(q + 2) \cdots (q + n).$$

It is our intention to apply Dirichlet's theorem to the progression $\{am + q\}$, and therefore we need to show that a and q are relatively prime.

Clearly $a > q$, and since q is a prime number, if a and q are not relatively prime, then q must divide a. Now, if q divides a, then q must divide some factor $(q \pm i)$ of a. But this would require q to divide i, and with $q > n \geq i$, this is impossible, and we conclude that q and a must be relatively prime.

Hence, by Dirichlet's theorem, $\{am + q\}$ contains an infinity of primes. Let $p = am + q$ be a prime in this progression for which $m > 0$ (making $p > q$). Then the n consecutive integers on either side of p are easily seen to be composite. They are the numbers

$$(am + q) - n, \ldots, (am + q) - 1, (am + q) + 1, \ldots, (am + q) + n,$$

that is, the numbers $\{am + (q \pm i)\}, i = 1, 2, \ldots, n$.

Now, since each $(q \pm i)$ divides a, it is a divisor of $am + (q \pm i)$, clearly a proper divisor since it is obviously less than $am + (q \pm i)$. And since $q - i \geq q - n \geq 2$, each $(q \pm i) > 1$, implying that each of the integers $am + (q \pm i)$ is indeed composite.

3. Next let us look at two variants of well known methods of proof that were kindly sent to me by Professor Nathan Mendelsohn of the University of Manitoba.

(a) The method of induction generally establishes a proposition $P(m)$ for $m \geq k$ by first showing the validity of $P(k)$ and then establishing that $P(n+1)$ holds when all the previous cases $P(k), P(k+1), \ldots, P(n)$, are assumed:

> thus $P(k+1)$ holds since $P(k)$ does, and then $P(k+2)$ holds because both $P(k)$ and $P(k + 1)$ do, and so on.

It often is enough to assume only $P(n)$ in order to prove $P(n+1)$, and it is easy to forget that we may assume any number of initial cases for this task. No matter how many initial cases might be used, one never needs to prove independently more than just the single initial case $P(k)$. Now let us look at a simple variant of induction.

Sometimes it is easier to establish $P(n+2)$ than $P(n+1)$ from the assumption of $P(n)$. In this case, $P(k)$ yields only $P(k+2)$, $P(k+4)$, $P(k+6)$, ... Hence it is necessary also to give an independent proof of $P(k+1)$ in order to obtain the cases $P(k+1)$, $P(k+3)$, $P(k+5)$, ... Let us consider two examples.

(i) Suppose we would like to prove that, for all $n = 1, 2, 3, \ldots$, there exists a solution (x, y, z) in positive integers of the equation

$$x^2 + y^2 = z^n.$$

The assumption of a solution (x_1, y_1, z_1) for any value of n,

$$x_1^2 + y_1^2 = z_1^n,$$

immediately establishes the result for $n + 2$:

multiplying by z_1^2 gives $(z_1 x_1)^2 + (z_1 y_1)^2 = z_1^{n+2}$.

Since both the cases $n = 1$ and $n = 2$ are obviously valid, the conclusion follows for all n by *leapfrog* induction.

(ii) Next let us prove the formula

$$1^2 - 2^2 + 3^2 - 4^2 + - \cdots + (-1)^{n-1} n^2$$
$$= (-1)^{n-1}(1 + 2 + 3 + \cdots + n),$$

which is easily verified for $n = 1$ and $n = 2$. The left side of the equation for the case $n + 2$ is

$$\left(1^2 - 2^2 + 3^2 - 4^2 + - \cdots + (-1)^{n-1} n^2\right)$$
$$+ (-1)^n (n+1)^2 + (-1)^{n+1} (n+2)^2.$$

Assuming the formula for n, and substituting for the large bracket, we obtain

$$(-1)^{n-1}(1 + 2 + 3 + \cdots + n) + (-1)^n (n+1)^2$$
$$+ (-1)^{n+1} (n+2)^2$$
$$= (-1)^{n-1}\big[(1 + 2 + 3 + \cdots + n) - (n^2 + 2n + 1)$$
$$+ (n^2 + 4n + 4)\big]$$

$$= (-1)^{n-1}(1 + 2 + 3 + \cdots + n + 2n + 3)$$
$$= (-1)^{n-1}\big[1 + 2 + 3 + \cdots + n + (n+1) + (n+2)\big],$$

which is equal to the right side of the formula for the case $n+2$ since $(-1)^{n-1} = (-1)^{n+1}$, completing the leapfrog induction.

(b) Our second variant concerns the so-called "method of infinite descent." This method is often used to show that a given equation has no solution in positive integers. The idea is to show that the assumption of a solution leads to the existence of another solution, also in positive integers, in which a specific one of the variables has a smaller value. To use a famous example, if (x_1, y_1, z_1) is a solution in positive integers of the equation

$$x^4 + y^4 = z^2,$$

then it can be shown that there exists a solution (x_2, y_2, z_2) in positive integers in which $z_1 > z_2$. Repeated use of the same argument establishes an unending decreasing sequence of positive integers $z_1 > z_2 > z_3 > \cdots$, which is clearly impossible, and the desired conclusion follows by contradiction.

Note that it is the existence of the decreasing infinite sequence of positive integers that gives the contradiction. If an initial assumption leads to a positive integer z_1 which, by some argument, leads to a smaller positive integer z_2 which similarly can be shown to lead to a still smaller positive integer z_3, and so on indefinitely, it doesn't matter where the integers z_i come from; they might all arise as solutions of a given equation or as solutions of an endless sequence of different equations.

Consider the problem of proving there is no solution in positive integers of the equation

$$x^2 + y^2 + z^2 = 2xyz.$$

Suppose that a solution (x_1, y_1, z_1) exists:

$$x_1^2 + y_1^2 + z_1^2 = 2x_1y_1z_1.$$

First we note that this implies each of x_1, y_1, z_1 must be an even integer:

if exactly one, or all three, of x_1, y_1, z_1 is odd, then we have the contradiction that the left side, $x^2 + y^2 + z^2$, is odd while the right side is even;

suppose exactly two of x_1, y_1, z_1 are odd: for definiteness, suppose $x_1 = 2a + 1$ and $y_1 = 2b + 1$, and $z_1 = 2c$. Then the equation is

$$4a^2 + 4a + 1 + 4b^2 + 4b + 1 + 4c^2 = 2(2a+1)(2b+1)2c$$

in which the right side is divisible by 4 while the left is not.

Now, since (x_1, y_1, z_1) is a solution, by dividing the equation by 4, we get

$$\left(\frac{x_1}{2}\right)^2 + \left(\frac{y_1}{2}\right)^2 + \left(\frac{z_1}{2}\right)^2 = 2^2 \left(\frac{x_1}{2}\right)\left(\frac{y_1}{2}\right)\left(\frac{z_1}{2}\right),$$

implying that $(\frac{x_1}{2}, \frac{y_1}{2}, \frac{z_1}{2})$ is a solution of the equation

$$x^2 + y^2 + z^2 = 2^2 xyz,$$

a solution in positive integers because each of x_1, y_1, z_1 is an even integer. Furthermore, it is obvious that $\frac{z_1}{2} < z_1$. Labelling this solution (x_2, y_2, z_2), successive applications of this line of reasoning give, for all n, a positive integer solution (x_n, y_n, z_n) of the equation

$$x^2 + y^2 = 2^n xyz,$$

where $z_n < z_{n-1}$ and the desired impossible sequence $z_1 > z_2 > z_3 > \ldots$ is established.

SECTION 21
Three Geometry Problems

1. The circle $A(4)$ (i.e., with centre A and radius 4) touches the circle $B(5)$ internally at C (Figure 1). D and E are arbitrary points on $A(4)$ and $B(5)$, respectively. Without using calculus, determine the maximum area of triangle CDE.

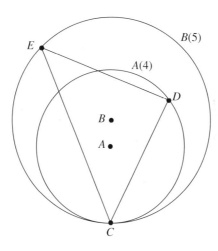

FIGURE 1

First observe that A, B, and C are collinear because the line joining the centers of touching circles goes through their point of contact.

Clearly it suffices to consider positions of D just on one side of the axis of symmetry CAB. Suppose that D is fixed on $A(4)$ and that $\angle ACD = \theta$ (Figure 2); also, suppose CAB meets $A(4)$ again at F.

Now, for a **fixed chord** CD, the greatest triangle CDE is the one having the greatest altitude, and therefore will have vertex E at the point of contact of the farther of the two tangents to $B(5)$ which are parallel to CD (think of CD being translated across the the figure, over the centre B, increasing the

169

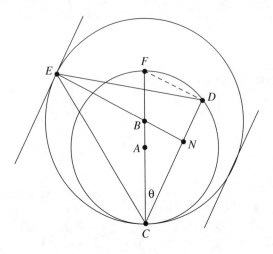

FIGURE 2

altitude all the while, to its final contact with $B(5)$ at E; clearly, translating in the opposite direction to the other tangent in this direction yields a smaller altitude). The radius BE is perpendicular to this tangent, implying that it is also perpendicular to CD. Hence EB extended gives the greatest altitude EN, and its length is

$$EN = 5 + BN = 5 + 5\sin\theta \quad \text{(from right triangle } BCN\text{)},$$
$$= 5(1 + \sin\theta).$$

Since CF is a diameter of $A(4)$, we have from right triangle CDF that

$$CD = CF\cos\theta = 8\cos\theta.$$

Hence

$$\triangle CDE = \tfrac{1}{2} CD \cdot EN = 20\cos\theta(1 + \sin\theta).$$

Thus, as D and θ vary, $\triangle CDE$ attains its overall greatest area when $K = \cos\theta(1 + \sin\theta)$ is greatest, and a straightforward completion of the solution is now provided by calculus. However, it is interesting to consider how one might proceed without calling on such a powerful technique.

Clearly K and K^2 attain their maximum values for the same value of θ. Now,

$$K^2 = \cos^2\theta(1 + \sin\theta)^2 = (1 - \sin^2\theta)(1 + \sin\theta)^2,$$

and for $\theta = (0, 30°, 45°, 60°, 90°)$, K^2 is approximately equal to

$$\left(1, \frac{27}{16}, \frac{24}{16}, \frac{14}{16}, 0\right),$$

respectively. This suggests that the maximum occurs in the vicinity of 30°, possibly at 30° itself. In the hope of this possibility,

let $\sin \theta = x + \frac{1}{2}$, where $|x| \leq \frac{1}{2}$ in order to place $\sin \theta$ in $[0, 1]$.

It would be much to our liking, then, to find that K^2 attains its maximum at $x = 0$. Now,

$$K^2 = \left[1 - \left(x + \frac{1}{2}\right)^2\right]\left(1 + x + \frac{1}{2}\right)^2$$

$$= \left(\frac{3}{4} - x - x^2\right)\left(\frac{9}{4} + 3x + x^2\right)$$

$$= \frac{27}{16} + \frac{9}{4}x + \frac{3}{4}x^2 - \frac{9}{4}x - 3x^2 - x^3 - \frac{9}{4}x^2 - 3x^3 - x^4$$

$$= \frac{27}{16} - \frac{9}{2}x^2 - 4x^3 - x^4$$

$$= \frac{27}{16} - x^2\left(\frac{9}{2} + 4x + x^2\right)$$

$$= \frac{27}{16} - x^2\left[(x + 2)^2 + \frac{1}{2}\right]$$

$$\leq \frac{27}{16}, \quad \text{with equality if and only if } x = 0.$$

Therefore the maximum value of K^2 is $\frac{27}{16}$, occurring at $x = 0$.

Hence K attains its maximum for $\sin \theta = \frac{1}{2}$, $\theta = 30°$, and the overall greatest area of $\triangle CDE$ is

$$20(\text{maximum } K) = 20\sqrt{\frac{27}{16}} = 20\left(\frac{3}{4}\sqrt{3}\right) = 15\sqrt{3}.$$

2. In parallelogram $ABCD$, E is an arbitrary point on AB (Figure 3). F is on AD and $BF = ED$. If P is the point of intersection of ED and BF, prove that PC bisects $\angle BPD$.

Clearly, the area of each of the triangles ECD and BCF is one-half the parallelogram, and therefore, they are equal. Since they have equal bases ED and BF, their altitudes from C are also equal. Thus C is equidistant from the arms of $\angle BPD$, and accordingly lies on the bisector of the angle.

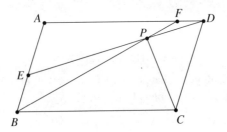

FIGURE 3

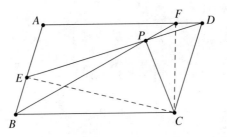

FIGURE 4

3. The square $ABCD$ is enclosed in a circle with center B and radius BD as in Figure 5. The line through A parallel to BD meets the circle at E. How big is $\angle ABE$?

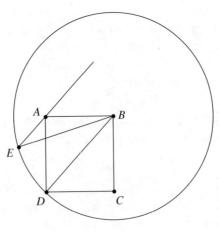

FIGURE 5

Three Geometry Problems

This is problem 81–13, on page 12 of the September 1981 issue of *The Ontario Secondary School Mathematics Bulletin* (published by the University of Waterloo).

(a) First a lovely synthetic solution, due to Ontario high school students Todd Cardno (Cardinal Newman H. S., Stoney Creek) and Peter Fowler (Grey Highlands S. S., Flesherton).

Draw *BF* perpendicular to *EA* (Figure 6). Since the diagonals of a square are perpendicular, *AC* is perpendicular to *BD*, and thus also to *AE*. Hence, in quadrilateral *FAGB*, the angles at *F*, *A*, and *G* are right angles, making it a rectangle, in fact, a square since $AG = BG$. Since radii *BD* and *BE* are equal, we have

$$FB = AG = \tfrac{1}{2}AC = \tfrac{1}{2}BD = \tfrac{1}{2}EB.$$

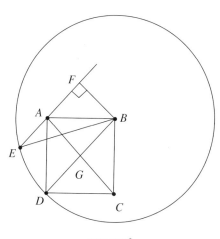

FIGURE 6

Thus right triangle *EBF* is a 30°–60°–90° triangle with $\angle EBF = 60°$.

Now, the diagonals of a square bisect the right angles through which they pass, and so, in square *FAGB*, $\angle ABF = 45°$, making

$$\angle ABE = \angle EBF - \angle ABF = 60° - 45° = 15°.$$

(b) Now a trigonometric approach.

Let $\angle ABE = x$ and let *AB* be the unit of length (Figure 7). Then $BD = \sqrt{2} = BE$. Since *EA* and *BD* are parallel, equal alternate angles yield

$$\angle HAB = \angle ABD = 45°,$$

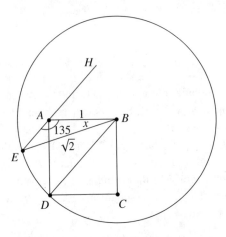

FIGURE 7

making
$$\angle EAB = 135°.$$

In $\triangle ABE$, then, $\angle E = 45° - x$, and by the law of sines
$$\frac{\sin(45° - x)}{1} = \frac{\sin 135°}{\sqrt{2}},$$
that is,
$$\sin(45° - x) = \frac{\sin 45°}{\sqrt{2}} = \frac{1}{2}.$$

Hence
$$45° - x = 30°,$$
and
$$x = 15°.$$

Comment: Even though solution (b) is easier, solution (a) takes pride of place. The construction of line BF is a brilliant idea which leads to a beautiful argument!

SECTION 22
Three Problems From The 1990 Balkan Olympiad

1. (Proposed by Greece) *

Let $a_1 = 1, a_2 = 3$, and for $n > 2$,
$$a_n = (n + 1)a_{n-1} - na_{n-2}.$$
Find all values of n for which a_n is divisible by 11.

Observe that
$$a_{10} = 11a_9 - 10a_8$$
and
$$a_{11} = 12a_{10} - 11a_9.$$
Thus, if a_8 is divisible by 11, so is a_{10} and then also a_{11}.

Now, the given recursion implies that if two consecutive terms are divisible by 11, so are all later terms. Thus, if a_8 is divisible by 11, so are all a_n for $n \geq 10$. Consequently, we need to check a_8. Simple calculations yield the first nine terms to be

1, 3, 9, 33, 153, 873, 5913, 46233, 409113.

Thus $a_8 = 46233$ is divisible by 11. Noting also that $a_4 = 33$ is divisible by 11, but no other terms less than a_{10} except a_8, it follows that a_n is divisible by 11 for $n = 4, 8$, and all $n \geq 10$.

2. (Proposed by Romania)

Find the minimum number of elements in a set A such that there exists a function $f : N \to A$ (from the positive integers N to A) having the property that $f(i) \neq f(j)$ whenever $|i - j|$ is a prime number.

175

If $f(n)$ is the remainder 0, 1, 2, or 3, that is obtained by dividing n by 4, then $f(i)$ and $f(j)$ can be equal only when $|i-j|$ is a multiple of 4, implying $f(i)$ and $f(j)$ are unequal when $|i-j|$ is a prime number. Thus four elements suffice for the set A.

On the other hand, since each two of the integers 1, 3, 6, 8 differ by a prime number, no two of the values $f(1)$, $f(3)$, $f(6)$, $f(8)$ can be the same, implying that A must have at least four members.

Thus the minimum number of elements in A is four.

3. (Proposed by Yugoslavia)

$A_1 B_1 C_1$ is the orthic triangle of an acute-angled, non-equilateral triangle ABC (Figure 1). The incircle of $\triangle A_1 B_1 C_1$ touches its sides at A_2, B_2, C_2. Prove that triangles ABC and $A_2 B_2 C_2$ have a common Euler line.

(Recall that the orthocenter of a triangle is the point of concurrency of the altitudes, that the orthic triangle is determined by the feet of the altitudes, and that the Euler line is the line joining the orthocenter and the circumcenter.)

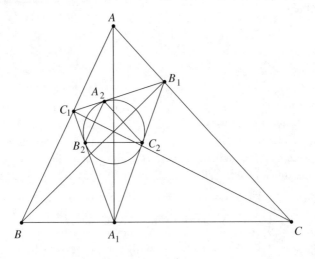

FIGURE 1

Our task, then, is to show that the two orthocenters and two circumcenters of triangles ABC and $A_2 B_2 C_2$ all lie on the same straight line.

Let H be the orthocenter of $\triangle ABC$ and O its circumcenter (Figure 2). Now, it is well known that the altitudes bisect the angles of the orthic triangle (this is proved in the Appendix). Thus H is the incenter of $\triangle A_1 B_1 C_1$. It fol-

Three Problems From The 1990 Balkan Olympiad

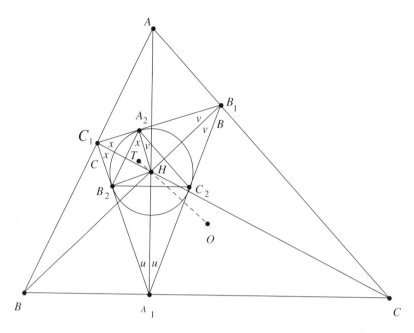

FIGURE 2

lows that the three inradii HA_2, HB_2, HC_2 are equal, which implies that H is also the circumcenter of $\triangle A_2 B_2 C_2$. Also, the radii HA_2, HB_2, and HC_2 are respectively perpendicular to the sides of $\triangle A_1 B_1 C_1$.

Thus H serves double duty as the orthocenter of $\triangle ABC$ and the circumcenter of $\triangle A_2 B_2 C_2$, and it remains to show that the orthocenter T of $\triangle A_2 B_2 C_2$, the point H, and the circumcenter O of $\triangle ABC$ are collinear.

It pays to draw a fairly accurate figure; even a freehand diagram was enough to suggest that the sides of triangles ABC and $A_2 B_2 C_2$ might be respectively parallel. It turned out to be easy to prove. For example, for sides AB and $A_2 B_2$:

the right angles at A_2 and B_2 make quadrilateral $C_1 A_2 H B_2$ cyclic, in whose circumcircle $B_2 H$ subtends the same angle x at C_1 and A_2. Since HC_1 bisects angle C_1, then angle $A_2 C_1 H$ also equals x. Thus the alternate angles $AC_1 A_2$ and $C_1 A_2 B_2$ are each the complement of x, and AB and $A_2 B_2$ are parallel.

Having their respective sides parallel, triangles ABC and $A_2 B_2 C_2$ are similar, and moreover, they are homothetic; that is to say, there is a homothecy (i.e., a dilatation) which takes the one triangle into the other (this is proved in the Appendix that immediately follows this solution).

The desired conclusion is now easily reached by observing that this homothecy would take the parts of $\triangle ABC$ into the corresponding parts of $\triangle A_2B_2C_2$, in particular it would take H to T and O to H. That is to say, HO is taken to TH. But homothecies do not change the direction of segments, and so TH and HO are parallel. Having the point H in common, it follows that T, H, and O are indeed collinear.

Appendix

(a) To prove the angles u at A_1 in Figure 2 are equal.

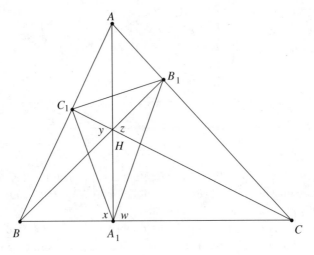

FIGURE 3

The right angles at C_1 and A_1 make C_1HA_1B cyclic in which angles x and y on chord BC_1 are equal. Similarly, B_1HA_1C is cyclic and the angles z and w are equal.

Since the vertically opposite angles y and z are equal, it follows that x and w are equal and also their complements (the angles u).

(b) To prove triangles ABC and $A_2B_2C_2$ are homothetic.

We already know that these triangles are similar. It remains to show that there exists a point from which one of them can be transformed into the other by a dilatation. Let BB_2 meet CC_2 at K (Figure 4), and let r denote the ratio of corresponding sides of triangle $A_2B_2C_2$ to the sides of triangle ABC. We will show that the dilatation with center K and ratio r takes triangle ABC to triangle $A_2B_2C_2$.

Three Problems From The 1990 Balkan Olympiad

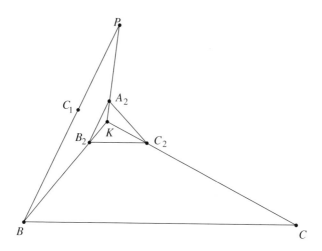

FIGURE 4

Let KA_2 meet BC_1 extended at P. First we show that P is the vertex A. Since B_2C_2 is parallel to BC, triangles KB_2C_2 and KBC are similar with corresponding sides in the ratio

$$\frac{B_2K}{BK} = \frac{B_2C_2}{BC} = r.$$

Since B_2A_2 is parallel to BA, it is also parallel to BC_1, and hence to BP, making triangles KB_2A_2 and KBP similar. Hence

$$\frac{B_2A_2}{BP} = \frac{B_2K}{BK} = r = \frac{B_2A_2}{BA},$$

and we have $BP = BA$, identifying P and A.

The similar triangles KB_2A_2 and KBP also give

$$\frac{A_2K}{PK} = r, \quad \text{that is,} \quad \frac{A_2K}{AK} = r,$$

and we have that the dilatation $K(r)$ takes A, B, C, to A_2, B_2, C_2, as desired.

SECTION 23
A Japanese "Fan" Problem

1. Introduction

During the period from about 1600 to 1850, a broad cross-section of the population of Japan, from farmer to samurai, indulged a passion for Euclidean geometry and made many amazing discoveries. Some 250 of these have been preserved in the wonderful book *Japanese Temple Geometry Problems* (in English) by Fukagawa and Pedoe (The Charles Babbage Research Centre, Winnipeg, Canada, 1989).

Figures constructed in a fan (a sector of a circle) became a popular topic of investigation. The problem we shall consider was first published in Japanese in 1833 by Shin Oh Jiku in Mathematics Summary (Sangaku Teiyo, 3 vols.). The problem was solved by Shinko Takeda, the son of the prominent mathematician and teacher Shingen Takeda.

Problems were often posed by presenting a figure accompanied by a minimum of description. In Figure 1, it is given that there are circles of just three

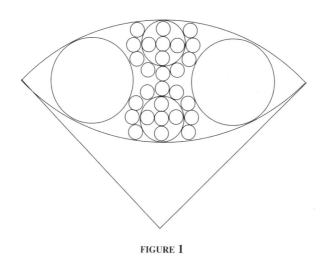

FIGURE 1

181

sizes—small, medium, and large (the tangencies and symmetries are assumed to be understood without comment). Then comes the problem—given that the radius of the fan is 10, how big are the large inscribed circles?

We follow the solution given by S. Tanaka, A. Kubo, and N. Sato in their paper "The Fan Problem and Its History" in the Japanese mathematics journal *Gurukula Kangri Jijñana Patrika Aryabhata*, Vol. 1, 1998, No. 1, 15–24. I am most grateful to S. Tanaka for including a reprint of their paper in our correspondence. The answer is

$$\text{the diameter} = 10 \cdot \frac{164011914 - 189728\sqrt{13074}}{327787557}$$
$$= 4.3417789 \quad \text{to seven places of decimals.}$$

2. The Solution

Let the center of the fan be A, its radius $a(=10)$, and the radius of the small circles b; also let the desired radius of the large inscribed circles be x.

We begin with the observation that the "cross" of small circles in a circle of medium size implies the radius of a medium circle is $3b$.

Next consider $\triangle ABC$ in the section of the configuration shown in Figure 2. Because of the various tangencies, it is clear that

$$AB = a - b, \quad AC = a - 3b, \quad \text{and} \quad BC = b + 3b = 4b.$$

Then, by the law of cosines,

$$AB^2 = BC^2 + AC^2 - 2 \cdot BC \cdot AC \cdot \cos \angle ACB,$$
$$a^2 - 2ab + b^2 = 16b^2 + (a^2 - 6ab + 9b^2)$$
$$- (8ab - 24b^2) \cdot \cos \angle ACB. \tag{1}$$

In order to calculate $\cos \angle ACB$, we observe in $\triangle CST$ that $CS = CT = 4b$ and $ST = 2b$; moreover, in Figure 2, since $ST = TE = EK = KB$, we have $\angle SCT = \frac{1}{4}\angle ACB$. Letting $\angle SCT = \theta$, then $\angle ACB = 4\theta$, and the law of cosines applied to $\triangle CST$ yields

$$\cos \theta = \frac{16b^2 + 16b^2 - 4b^2}{2 \cdot 4b \cdot 4b} = \frac{7}{8}.$$

Hence

$$\cos 2\theta = 2\cos^2 \theta - 1 = \frac{17}{32},$$

A Japanese "Fan" Problem

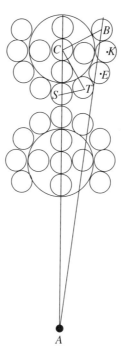

FIGURE 2

and

$$\cos \angle ACB = \cos 4\theta = 2\left(\frac{17}{32}\right)^2 - 1 = -\frac{223}{512}.$$

Therefore equation (1) becomes

$$a^2 - 2ab + b^2 = 16b^2 + (a^2 - 6ab + 9b^2) + (8ab - 24b^2) \cdot \frac{223}{512},$$

which reduces to

$$b = \frac{11}{289} \cdot a = \frac{110}{289} \quad (\text{recall } a = 10).$$

In the part of the configuration shown in Figure 3, let the center of the large circle be D and, as in Figure 2, let the center of the small circle between the large and medium circles be E. Let AC cross the line of centers of the two large inscribed circles at H, and from E let perpendiculars be drawn to meet AC at F and HD at G (it is not known that F is the center of a small circle).

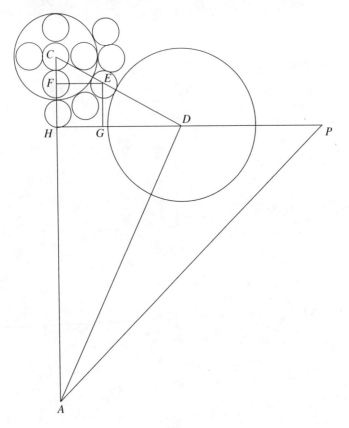

FIGURE 3

Then, in right triangle FEC, we have $CE = 4b$ and $\angle FCE = 2\theta$ (recall from Figure 2 that $\theta = \angle SCT$). Hence

$$CF = CE \cos 2\theta = 4b \cdot \frac{17}{32} = \frac{17}{8}b,$$

and, in rectangle $FHGE$,

$$HG = EF = CE \sin 2\theta = 4b\sqrt{1 - \left(\frac{17}{32}\right)^2} = \frac{7\sqrt{15}}{8}b.$$

It follows that

$$EG = FH = CH - CF = 5b - \frac{17}{8}b = \frac{23}{8}b.$$

Then, recalling that the radius of the circle with center D is x, the Pythagorean theorem applied to $\triangle GED$ gives

$$GD = \sqrt{DE^2 - EG^2} = \sqrt{(b+x)^2 - \frac{529}{64}b^2} = \sqrt{x^2 + 2bx - \frac{465}{64}b^2}.$$

Thus

$$HD = HG + GD = FE + GD = \frac{7\sqrt{15}}{8}b + \sqrt{x^2 + 2bx - \frac{465}{64}b^2},$$

and, applying the theorem of Pythagoras to $\triangle ADH$, we obtain

$$AD^2 = HD^2 + AH^2,$$

that is,

$$(a-x)^2 = \left(\frac{7\sqrt{15}}{8}b + \sqrt{x^2 + 2bx - \frac{465}{64}b^2}\right)^2 + (a-8b)^2.$$

Finally, substituting $a = 10$ and $b = \frac{110}{289}$ and simplifying, this yields the required diameter to be

$$2x = 10 \cdot \frac{164011914 - 189728\sqrt{13074}}{327787557}$$
$$= 4.3417789 \quad \text{to seven decimal places.}$$

It is remarkable that Shinko Takeda, the original solver, calculated the diameter to this degree of accuracy by "Soroban," a sort of abacus.

Finally, we observe in Figure 3 that

$$\cos \angle HAP = \frac{HA}{AP} = \frac{a-8b}{a}$$
$$= 1 - 8 \cdot \frac{b}{a} = 1 - 8 \cdot \frac{11}{289} = \frac{201}{289}.$$

Thus the angle at the center of the fan is

$$2\angle HAP = 2\arccos\frac{201}{289},$$

an angle of approximately 91.9 degrees.

SECTION 24
Slicing a Doughnut

This section is based on the splendid classroom capsule *Doughnut Slicing* by Wolf von Rönik, Portfolio Analytics Inc., Chicago, Illinois, which appeared in *The College Mathematics Journal*, Vol. 28, November 1997, pages 381–382.

Clearly a doughnut is sliced in half by any plane through its center, that is, the center of the hole in the middle. Since we shall be concerned only with the surface of the doughnut, it would really be better to call it an inner-tube or, to be mathematically precise, a torus.

Anyway, imagine a doughnut lying on a table (Figure 1). Let M be a horizontal straight line through the center O and, in the vertical plane through O which is perpendicular to M, let N be the straight line through O that just rests on the surface of the doughnut. Consider the tilted plane π that is determined by M and N. The question is

"How does π slice through the doughnut? Can you picture their intersection?"

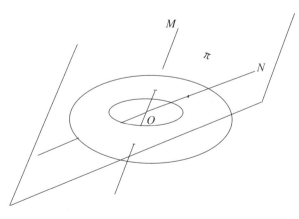

FIGURE 1

187

For most of us it is far beyond our powers of visualization to imagine what it would look like. Thus it might come as a pleasant surprise to learn that the intersection is a pair of circles! In view of the symmetry, it is hardly a further surprise that the circles are the same size and are located symmetrically with respect to the center. As we shall see, the proof of all these things is a nice straightforward application of elementary analytic geometry.

Suppose the doughnut is generated by revolving about the z-axis the circle whose center in the xz-plane is $(R, 0)$ and whose radius is r (Figure 2). Of course, in order to have a hole in the middle, R must be bigger than r.

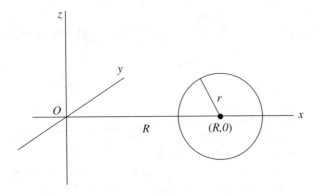

FIGURE 2

Now consider any point $P(x, y, z)$ on the doughnut (Figure 3). Let C be the center of the generating circle as it passes through P and let Q be the foot of the perpendicular from P to the xy-plane. Since P is directly above or below the line OC, as the case might be, then Q lies on the line OC, $PQ = z$, and

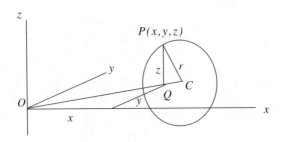

FIGURE 3

Q is the point (x, y) in the xy-plane. Hence $OQ = \sqrt{x^2 + y^2}$; also, $OC = R$ and $PC = r$. Hence $|QC| = |R - \sqrt{x^2 + y^2}|$, and from triangle PQC, which is right-angled at Q, we obtain the equation of the doughnut to be

$$\left(R - \sqrt{x^2 + y^2}\right)^2 + z^2 = r^2.$$

Clearly, the origin O is the center of the doughnut. Referring again to Figure 1, if we let M be the y-axis, then the line N lies in the xz-plane and is tangent to the initial position of the generating circle (Figure 4). If N makes an angle θ with the x-axis, then

$$\sin\theta = \frac{r}{R}.$$

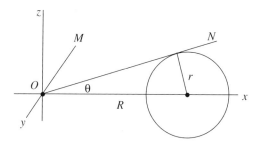

FIGURE 4

In the xz-plane, the equation of the line N is $\frac{z}{x} = \tan\theta$ (Figure 5). Now, since the cutting plane π is determined by N and the y-axis M, a straight line parallel to the y-axis through a point on N lies entirely in the plane π. Thus, if (x, z) is a point on N, the point $P(x, y, z)$ lies on π for all values of y. Hence,

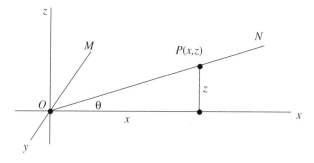

FIGURE 5

in order for $P(x, y, z)$ to lie on π, it is only necessary that $z/x = \tan\theta$, and it follows that the equation of π is simply

$$z = x \tan\theta.$$

Hence we seek the intersection of the surfaces

$$\left(R - \sqrt{x^2 + y^2}\right)^2 + z^2 = r^2 \quad \text{and} \quad z = x \tan\theta.$$

This turns out to be very awkward algebraically, and so let us try to make things easier for ourselves.

Accordingly, let the reference frame be rotated about the y-axis through the angle θ so that the xy-plane coincides with π (Figure 6). Then the equation of π becomes $z = 0$, and the intersection we seek is obtained by substituting zero for the new variable z' in the new equation of the doughnut. Since y-coordinates are unchanged by a rotation about the y-axis, the point (x, y, z) would now have coordinates (x', y', z') where $y' = y$ and x' and z' are given by the standard relations

$$x' = x \cos\theta + z \sin\theta,$$
$$z' = -x \sin\theta + z \cos\theta.$$

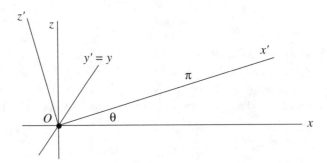

FIGURE 6

Since we have the equation of the doughnut in terms of x, y, z, we really need the inverse relations

$$x = x' \cos\theta - z' \sin\theta,$$
$$z = x' \sin\theta + z' \cos\theta.$$

Thus the equation of the doughnut referred to the new frame of reference is

$$\left(R - \sqrt{(x'\cos\theta - z'\sin\theta)^2 + y'^2}\right)^2 + (x'\sin\theta + z'\cos\theta)^2 = r^2,$$

and the desired intersection, given by $z' = 0$, is

$$\left(R - \sqrt{(x'\cos\theta)^2 + y'^2}\right)^2 + (x'\sin\theta)^2 = r^2.$$

At this point x' and y' have served their purpose. Hence, for easier reading, let us change back to just x and y. Thus we want to show that the equation

$$\left(R - \sqrt{(x\cos\theta)^2 + y^2}\right)^2 + (x\sin\theta)^2 = r^2$$

represents a pair of circles in the xy-plane. It remains only to go through the straightforward algebraic manipulations.

Expanding, we have

$$R^2 - 2R\sqrt{x^2\cos^2\theta + y^2} + (x^2\cos^2\theta + y^2) + x^2\sin^2\theta = r^2,$$

$$x^2 + y^2 + R^2 - r^2 = 2R\sqrt{x^2\cos^2\theta + y^2}.$$

Squaring gives $[(x^2 + y^2) + (R^2 - r^2)]^2 = 4R^2 x^2 \cos^2\theta + 4R^2 y^2$,

$$(x^2 + y^2)^2 + 2(x^2 + y^2)(R^2 - r^2) + (R^2 - r^2)^2 = 4R^2 x^2 \cos^2\theta + 4R^2 y^2.$$

Subtracting $4(x^2 + y^2)(R^2 - r^2)$ from each side yields

$$\left[(x^2 + y^2) - (R^2 - r^2)\right]^2 = 4R^2 x^2 \cos^2\theta + 4R^2 y^2 - 4(x^2 + y^2)(R^2 - r^2)$$
$$= 4R^2 x^2 (\cos^2\theta - 1) + 4r^2 x^2 + 4r^2 y^2$$
$$= 4R^2 x^2 (-\sin^2\theta) + 4r^2 x^2 + 4r^2 y^2.$$

Recalling that $\sin\theta = r/R$, then

$$\left[(x^2 + y^2) - (R^2 - r^2)\right]^2 = -4x^2 r^2 + 4r^2 x^2 + 4r^2 y^2,$$
$$\left[(x^2 + y^2) - (R^2 - r^2)\right]^2 = 4r^2 y^2.$$

Finally, transposing and factoring the difference of squares we obtain

$$\left\{\left[(x^2 + y^2) - (R^2 - r^2)\right] - 2ry\right\}\left\{\left[(x^2 + y^2) - (R^2 - r^2)\right] + 2ry\right\} = 0,$$

that is,

$$\left[x^2 + (y - r)^2 - R^2\right]\left[x^2 + (y + r)^2 - R^2\right] = 0,$$

giving

$$x^2 + (y - r)^2 = R^2 \quad \text{or} \quad x^2 + (y + r)^2 = R^2,$$

which is a pair of circles of radius R with centers at $(0, \pm r)$ in the xy-plane.

Even with the knowledge of the size and positions of these circles, you might still find it a stiff challenge to picture them clearly in your mind's eye. In Figure 7, we can imagine the intersection being formed by considering the cutting ray ON as it rotates counterclockwise in π about the center O.

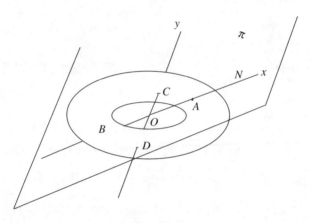

FIGURE 7

Beginning tangentially at its highest position at A on the right side, it slices through the surface along the two curves of a crescent (Figure 8). Proceeding downward around π, the crescent broadens as its inner curve makes

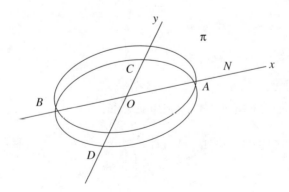

FIGURE 8

its way around the hole, through C, while its outer curve goes around the very outside of the doughnut at the back. After one-quarter of the way around, the crescent begins to narrow until at halfway it arrives at its lowest point B, where it is again tangent to the surface and the crescent closes. The second half of the rotation yields an identical crescent running upward from B to A.

Now it is not difficult to imagine that the inner arc ACB of the first crescent goes with the outer arc BDA of the second crescent to form one of the circles, and the other two arcs form an identical second circle which is symmetric with respect to the center O.

SECTION 25
A Problem from the 1980 Tournament of the Towns

This problem, due to Agnis Andjans of Riga, Latvia, appeared on the very first Tournament of the Towns in 1980. The brilliant and beautiful solution is due to Andy Liu, University of Alberta, Edmonton, Canada. We follow the account given in the exciting little book *Tournament of the Towns*, 1980–1984, edited by Peter Taylor (University of Canberra, Australia), published in 1993 by The Australian Mathematics Trust (Belconnen, ACT 2616, Australia).

The Problem

If each side of a convex quadrilateral $ABCD$ is divided into n equal parts and the corresponding points in opposite sides are joined, the quadrilateral is partitioned into n^2 sub-quadrilaterals which are generally of different shapes and sizes. However, prove that, if n of these sub-quadrilaterals are chosen along any transversal of the partition, that is, with no two in the same row or in the same column, then the sum of their areas is always $\frac{1}{n}ABCD$. (In Figure 1: $s + t + u + v = \frac{1}{4}ABCD$.)

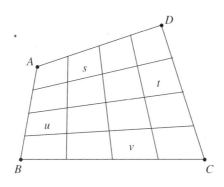

FIGURE 1

1. The first step is to show that each segment across $ABCD$ is also divided into n equal pieces by the segments which cross it. This is an easy corollary of the following property (Figure 2):

if P and Q divide the opposite sides AB and CD in the same ratio $a : b$ and R and S divide the other pair of opposite sides BC and CD in the same ratio $c : d$, then PQ and RS also divide each other in these same ratios:

$$\frac{PT}{TQ} = \frac{c}{d} \quad \text{and} \quad \frac{ST}{TR} = \frac{a}{b}. \tag{1}$$

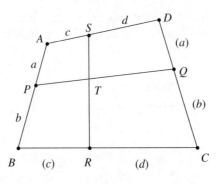

FIGURE 2

The result is trivial if $ABCD$ is a parallelogram. Suppose, then, that $ABCD$ is not a parallelogram (Figure 3). In this case complete parallelogram $ABED$ and partition it into four sub-parallelograms with lines PG and SF parallel to

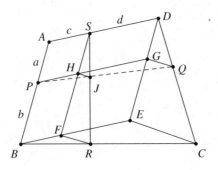

FIGURE 3

A Problem from the 1980 Tournament of the Towns   197

its sides. Suppose *PG* and *SF* cross at *H* and that *HJ* is drawn parallel to *FR* to meet *SR* at *J*. Join *FR*, *EC*, and *GQ*, and suppose the actual lengths of *AP*, *PB*, *AS*, *SD* are, respectively, a, b, c, d.

We begin by showing that *J* lies on *PQ*. Since opposite sides of a parallelogram are equal, then $SH = a$ and $HF = b$, and we have

$$\frac{SH}{SF} = \frac{a}{a+b}.$$

Since *HJ* is parallel to *FR*, triangles *SHJ* and *SFR* are similar and

$$\frac{HJ}{FR} = \frac{SH}{SF} = \frac{a}{a+b}.$$

Also,

$$\frac{BF}{FE} = \frac{PH}{HG} = \frac{c}{d},$$

and since

$$\frac{BR}{RC} = \frac{c}{d},$$

then *F* and *R* divide *BE* and *BC* in the same ratio, implying *FR* is parallel to *EC* and that triangles *BFR* and *BCE* are similar. Hence

$$\frac{FR}{EC} = \frac{BF}{BE} = \frac{c}{c+d}.$$

Similarly, *GQ* is parallel to *EC* and

$$\frac{GQ}{EC} = \frac{a}{a+b}.$$

Thus, since *HJ*, *FR*, *EC*, and *GQ* are all parallel, *HJ* and *GQ* are parallel and their ratio is

$$\frac{HJ}{GQ} = \frac{HJ}{FR} \cdot \frac{FR}{EC} \cdot \frac{EC}{GQ} = \frac{a}{a+b} \cdot \frac{c}{c+d} \cdot \frac{a+b}{a} = \frac{c}{c+d}.$$

But

$$\frac{PH}{PG} = \frac{c}{c+d},$$

and since angles *PHJ* and *PGQ* are equal corresponding angles for the parallel lines *HJ* and *GQ*, it follows that triangles *PHJ* and *PGQ* are similar. Thus

$$\angle HPJ = \angle GPQ,$$

and we conclude that J lies on PQ, implying that J is in fact the point of intersection T of PQ and RS in Figure 2.

Finally, since HJ is parallel to FR, H and J divide SF and SR in the same ratio, namely $a : b$. Similarly, HJ is parallel to GQ and H and J divide PG and PQ in the same ratio $c : d$, establishing the desired results of line (1) above.

Corollary. *In Figure 4, WZ joins corresponding points on AB and CD, and KL and MN are* consecutive *segments which join corresponding points on AD and BC. It is easy to see that* $XY = \frac{1}{n}WZ$:

for some positive integer t we have

$$AK = \frac{t}{n}AD \quad \text{and} \quad AM = \frac{t+1}{n}AD,$$

and since the ratios on AD are induced on WZ, then

$$WX = \frac{t}{n}WZ \quad \text{and} \quad WY = \frac{t+1}{n}WZ.$$

Hence

$$XY = WY - WX = \frac{1}{n}WZ.$$

Thus each crossing segment is itself divided into n equal parts by the segments which cross it.

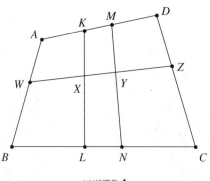

FIGURE 4

2. Next, consider a quadrilateral $EFGH$ that has been partitioned into mn subquadrilaterals by segments which join corresponding points that divide each side of one oppposite pair into n equal parts and each side of the other pair

A Problem from the 1980 Tournament of the Towns   199

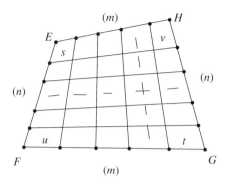

FIGURE 5

into m equal parts (Figure 5). We shall show that the sum of the areas of the quadrilaterals in opposite corners is the same for both pairs:

$$s + t = u + v.$$

Let us build up to the general $m \times n$ case. First consider the case of a 2×2 partition (Figure 6). Since P, Q, R, S are the midpoints of the sides, TP is a median of $\triangle ETF$ and bisects its area; similarly for TR, TQ, and TS. Hence each pair of quadrilaterals in opposite corners has total area $a + b + c + d$, and the claim is valid in this case.

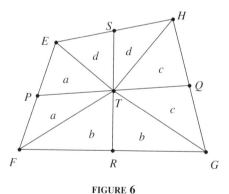

FIGURE 6

Next, consider a $2 \times m$ partition (Figure 7). Each segment along PQ is $\frac{1}{m}PQ$, and each cross-segment, like RS and TU, is bisected by PQ. Thus each consecutive pair of columns, $EFTU$, $SRVW$, $UTXY$, ..., comprises a quadrilateral that is decomposed into a 2×2 partition. Accordingly,

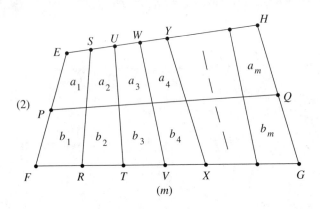

FIGURE 7

$$a_1 + b_2 = b_1 + a_2,$$
$$a_2 + b_3 = b_2 + a_3,$$
$$a_3 + b_4 = b_3 + a_4,$$
$$\vdots \quad \vdots \quad \vdots$$
$$a_{m-1} + b_m = b_{m-1} + a_m.$$

Adding up and subtracting the common terms from each side we obtain

$$a_1 + b_m = b_1 + a_m,$$

establishing the claim for a $2 \times m$ partition.

Finally, consider an $n \times m$ partition (Figure 8). Each pair of consecutive rows constitutes a $2 \times m$ partition and we have

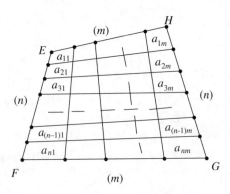

FIGURE 8

A Problem from the 1980 Tournament of the Towns   201

$$a_{11} + a_{2m} = a_{21} + a_{1m},$$
$$a_{21} + a_{3m} = a_{31} + a_{2m},$$
$$a_{31} + a_{4m} = a_{41} + a_{3m},$$
$$\vdots \quad \vdots \quad \vdots$$
$$a_{(n-1)1} + a_{nm} = a_{n1} + a_{(n-1)m},$$

and hence

$$a_{11} + a_{nm} = a_{1m} + a_{n1},$$

establishing the claim in all cases.

3. Now, an appropriate selection of n transversals covers $ABCD$: for example, in Figure 9, the five transversals marked a, b, c, d, e cover the 5×5 partition of $ABCD$. Finally, then, let us show that the total area of the quadrilaterals along a transversal is constant. From this it follows that each transversal has $\frac{1}{n}$th of the total area, as desired.

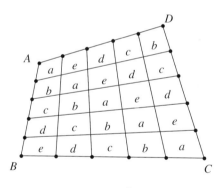

FIGURE 9

To this end, let us show that the sum of the areas along any transversal T is the same as the area along the transversal M determined by the main diagonal of the partition.

Suppose, then, that T is not the transversal M. In this case there is some quadrilateral Q on the main diagonal which does not belong to T (Figure 10). Now, T contains some quadrilateral X in Q's row and a quadrilateral Y in Q's column. Let Z be the intersection of X's column and Y's row. Then (Q, Z) and (X, Y) are pairs of opposite corners in a partitioned sub-quadrilateral of

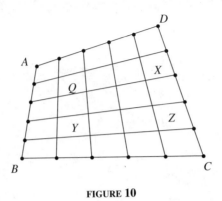

FIGURE 10

$ABCD$ and therefore the areas

$$X + Y = Q + Z.$$

We observe that Z, as well as Q, cannot belong to T, for T cannot have more than one quadrilateral in any row or column. Also, since Q lies on the main diagonal M, then X and Y, being respectively in the same row and column with Q, cannot lie on M. Thus, replacing X and Y with Q and Z does not take away from T's occupancy of the main diagonal but increases its holdings with the new quadrilateral Q (Z might or might not also lie on M). The result is a transversal T' with the same total area as T and which has at least one more quadrilateral on M than T does. Proceeding similarly until no empty places remain on M, it follows that T and M have the same area, and the proof is complete.

More Challenges

1. The positive integer 88 is 7 units greater than the nearest square below it (81) and 12 units less than the nearest square above it (100).

FIGURE 1

The product of 7 and 12 is 84, and $88 - 84 = 4$, a perfect square.

Now, every positive integer n lies in a half-closed interval $[k^2, (k+1)^2[$

$$k^2 \leq n < (k+1)^2.$$

Prove that, if $x = n - k^2$ and $y = (k+1)^2 - n$, then $n - xy$ is always a perfect square.

2. (From the *College Mathematics Journal*, Vol. 28, Nov. 1997, page 393)

What is the sum of the series $a_1 + a_2 + a_3 + \cdots$, where a_n is defined by Figure 2?

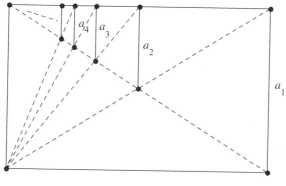

FIGURE 2

203

3. (Problem 617 from the problems section of *The College Mathematics Journal*, Vol. 29, January, 1998, page 66)

   Find all positive integer solutions (x, n) of $x^2 + 8 = 3^n$.

4. (From the 1998 Zimbabwe Olympiad)

   $\triangle ABC$ is right angled at A and AD is the altitude from A (Figure 3). $DC = 1$ and E is a point in AC such that $BE = EC = 1$. If $x = AE$, how long is x?

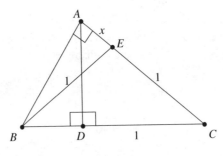

FIGURE 3

5. (From the 1998 Putnam Examination)

   Find the minimum value of

   $$\frac{\left(x + \frac{1}{x}\right)^6 - \left(x^6 + \frac{1}{x^6}\right) - 2}{\left(x + \frac{1}{x}\right)^3 + \left(x^3 + \frac{1}{x^3}\right)} \quad \text{for} \quad x > 0.$$

6. (From the 1998 Canadian Open Mathematics Challenge)

   $\triangle DXY$ is inscribed in rectangle $ABCD$ as shown in Figure 4. If the areas of triangles AXD, BXY, and DYC, respectively, are 5, 4, and 3 units, what is the area of $\triangle DXY$?

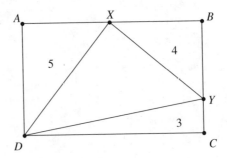

FIGURE 4

More Challenges

7. (A problem of Adam Brown)

   What is the area of the shaded region in Figure 5?

   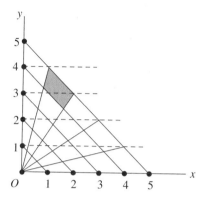

   FIGURE 5

8. (From the 1998 Canadian Open Mathematics Challenge)

   There are ten prizes, five A's, three B's, and two C's, placed in identical sealed envelopes for the top ten contestants in a mathematics contest. The prizes are awarded by allowing winners to select an envelope at random from those remaining. When the eighth contestant goes to select a prize, what is the probability that the remaining three prizes are one A, one B, and one C?

9. (Two exercises from *Quantum*, Vol. 7, July/Aug, 1997)

   (a) What is the minimum value of the function
   $$f(x) = \frac{x^2}{8} + x\cos x + \cos 2x?$$

   (b) Prove that the equation $x^3 - x - 3 = 0$ has a unique real root and that its value is greater than $\sqrt[5]{13}$.

10. (Problem B214 from *Quantum*, Vol. 8, Sept/Oct, 1997)

    Find the fraction $\frac{a}{b}$ with smallest denominator that lies between $\frac{97}{36}$ and $\frac{96}{35}$:
    $$\frac{97}{36} < \frac{a}{b} < \frac{96}{35}.$$

11. (Problem 637 from *The College Mathematics Journal*, Vol. 29, Nov., 1998)

The sequence $\{a_n\}$ is defined by
$$a_1 = \sqrt{2}, \quad \text{and for } n = 1, \quad a_{n+1} = \sqrt{2 + a_n}.$$

Determine
$$\lim_{n \to \infty} 4^n (2 - a_n).$$

12. (From the 1991 International Olympiad)

For any point P inside any $\triangle ABC$, show that at least one of the angles x, y, z does not exceed $30°$ (Figure 6).

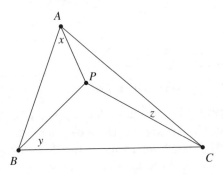

FIGURE 6

13. In Figure 7, $ABCD$ is a cyclic quadrilateral constructed on diameter AC as follows. B is chosen arbitrarily on one of the semicircular arcs AC. M is the midpoint of AB and the point P is taken on CM such that $MP = AM$. Extending BP then gives D on the circumference. Finally, CQ is drawn parallel to AP. Prove $BP = QD$ for all positions of B.

14. Seven re-phrased problems from the 1999 AIME (American Invitational Mathematics Examination)

    (i) A transformation of the first quadrant of the coordinate plane maps each point (x, y) to the point $(x', y') = (\sqrt{x}, \sqrt{y})$. The vertices of the quadrilateral $ABCD$ are $A = (900, 300)$, $B = (1800, 600)$, $C = (600, 1800)$, and $D = (300, 900)$. What is the area of the region enclosed by the image of $ABCD$?

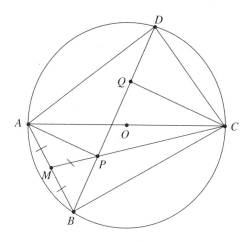

FIGURE 7

(ii) There is a set of 1000 switches, each of which has four positions, called A, B, C, D. When the position of any switch changes, it is only from A to B, from B to C, from C to D, or from D to A. Initially each switch is in position A. The switches are labeled with the 1000 different integers $2^x 3^y 5^z$, where each of x, y, z takes the values $0, 1, \ldots, 9$. Moreover, the switches are numbered switch 1, switch 2, ..., switch 1000.

At step i of a 1000-step process, the ith switch is advanced one step, and so are all the other switches whose labels divide the label on the ith switch. After step 1000 has been completed, how many switches will be in position A?

(iii) Let T be the set of ordered triples (x, y, z) of nonnegative real numbers that lie in the plane $x + y + z = 1$. Let us say that (x, y, z) *supports* (a, b, c) when exactly two of the following are true:

$$x \geq a, \quad y \geq b, \quad z \geq c.$$

Let S consist of those triples in T that support the point

$$P\left(\frac{1}{2}, \frac{1}{3}, \frac{1}{6}\right).$$

Determine the ratio

$$\frac{\text{area of } S}{\text{area of } T}.$$

(iv) Ten points in the plane are given, no three collinear. Four distinct segments joining pairs of these points are chosen at random, all such segments being equally likely. What is the probability that some three of the four chosen segments form a triangle whose vertices are among the ten given points?

(v) Find the acute angle θ, given that

$$\sum_{k=1}^{35} \sin 5k = \tan\theta,$$

where the angles are measured in degrees.

(vi) Forty teams play a tournament in which every team plays every other team exactly once. No ties occur, and each team has a 50% chance of winning any game it plays. Determine the probability that no two teams win the same number of games.

(vii) Consider the paper triangle whose vertices are $(0, 0)$, $(34, 0)$, and $(16, 24)$. The vertices of its medial triangle are the midpoints of its sides. A triangular pyramid is formed by folding the triangle along the sides of its medial triangle. What is the volume of this pyramid?

15. From William Dunham's wonderful book *Euler, The Master of Us All* (MAA, Dolciani Series, Vol. 22, 1999).

    Prove the remarkable discoveries of Jakob Bernoulli (1654–1705) that

    $$\text{(a)} \quad \sum_{k=1}^{\infty} \frac{k^2}{2^k} = 6 \quad \text{and} \quad \text{(b)} \quad \sum_{k=1}^{\infty} \frac{k^3}{2^k} = 26.$$

    It's amazing that the sum of such a series is a whole number!

Solutions To The Challenges

1. Every positive integer n lies in a half-closed interval $[k^2, (k+1)^2[$ for some positive integer k:

$$k^2 \leq n < (k+1)^2.$$

Prove that, if

$$x = n - k^2 \quad \text{and} \quad y = (k+1)^2 - n,$$

then

$$n - xy$$

is always a perfect square.

Clearly

$$xy = n(k+1)^2 - k^2(k+1)^2 - n^2 + nk^2,$$

and so

$$\begin{aligned}
n - xy &= n - n(k+1)^2 + k^2(k+1)^2 + n^2 - nk^2 \\
&= n - nk^2 - 2nk - n + k^4 + 2k^3 + k^2 + n^2 - nk^2 \\
&= k^4 + k^2 + n^2 + 2k^3 - 2nk^2 - 2nk \\
&= (k^2 + k - n)^2,
\end{aligned}$$

i.e., the result is always just the square of the difference between $k^2 + k$ and n.

2. What is the sum of the series $a_1 + a_2 + a_3 + \cdots$, where a_n is defined by the following figure?

209

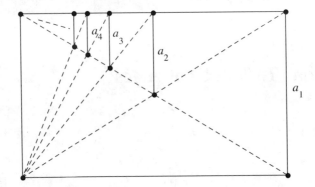

Let the rectangle have vertices at the origin O, $(a, 0)$, $(0, 1)$, and $(a, 1)$ (Figure 1). In this case $a_1 = \frac{1}{1}$ and clearly $a_2 = \frac{1}{2}a_1$, making $a_2 = \frac{1}{2}$. Now suppose $a_n = \frac{1}{n}$ for some $n \geq 1$. We shall see that $a_{n+1} = 1/(n+1)$.

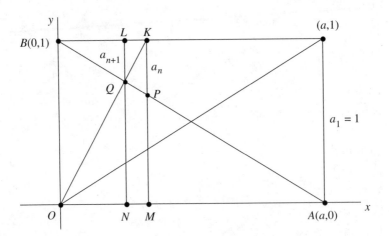

FIGURE 1

The equation of AB is easily found to be $x + ay = a$. Referring to Figure 1,

$$MP = 1 - \frac{1}{n} = \frac{n-1}{n},$$

and substituting this for y in the equation of AB, we get

$$OM = x = a\left(1 - \frac{n-1}{n}\right) = \frac{a}{n}.$$

Thus K is the point $(\frac{a}{n}, 1)$ and the equation of OK is

$$y = \frac{n}{a}x.$$

Solving for Q, we get

$$ay = nx = a - x \Rightarrow x = \frac{a}{n+1},$$

that is, $ON = a/(n+1)$. Substituting for x in the equation of AB, we get $NQ = y$ from

$$\frac{a}{n+1} + ay = a, \quad \text{i.e.,} \quad NQ = \frac{n}{n+1}.$$

Hence

$$QL = a_{n+1} = 1 - \frac{n}{n+1} = \frac{1}{n+1},$$

and it follows by induction that $a_n = \frac{1}{n}$ for all $n \geq 1$.

Thus the series is just a_1 times the harmonic series,

$$a_1\left(1 + \tfrac{1}{2} + \tfrac{1}{3} + \cdots\right),$$

which diverges.

3. Find all positive integer solutions (x, n) of $x^2 + 8 = 3^n$.

Two things strike one immediately: x must be odd and $x = 1, n = 2$ is a solution.

The key fact is that n must be even, making 3^n a square:

since x is odd, then $x^2 \equiv 1 \pmod{8}$, implying $3^n \equiv 1 \pmod{8}$; this is enough to show n is even, for $3^{2k+1} = 3 \cdot 9^k \equiv 3 \pmod{8}$.

Letting $n = 2k$, we have

$$3^{2k} - x^2 = 8,$$
$$(3^k + x)(3^k - x) = 8,$$

where the factors on the left are even and have the same sign. Since $3^k + x$ is positive, then both are. Being integers, the only possibility is $4 \cdot 2$, that is,

$$3^k + x = 4 \quad \text{and} \quad 3^k - x = 2.$$

Hence

$$2x = 2, x = 1, \quad \text{and} \quad k = 1.$$

Therefore $x = 1, n = 2$ is the only solution.

4. $\triangle ABC$ is right angled at A and AD is the altitude from A. $DC = 1$ and E is a point in AC such that $BE = EC = 1$. If $x = AE$, how long is x?

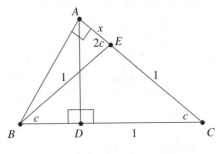

FIGURE 2

From $\triangle ADC$ we get

$$\cos C = \frac{1}{1+x}.$$

Now, in isosceles triangle BEC, the base angles are angle C, and so the exterior angle at E is

$$\angle AEB = 2C.$$

Thus, from $\triangle ABE$ we have

$$x = \cos 2C$$
$$= 2\cos^2 C - 1$$
$$= \frac{2}{(1+x)^2} - 1.$$

Hence

$$1 + x = \frac{2}{(1+x)^2},$$

and

$$x = \sqrt[3]{2} - 1.$$

Solutions To The Challenges

5. Find the minimum value of

$$\frac{\left(x+\frac{1}{x}\right)^6 - \left(x^6+\frac{1}{x^6}\right) - 2}{\left(x+\frac{1}{x}\right)^3 + \left(x^3+\frac{1}{x^3}\right)} \quad \text{for} \quad x > 0.$$

We were always taught that, when $x^n + 1/x^n$ appeared several times, let $x + \frac{1}{x} = y$. Accordingly,

$$y^3 = \left(x+\frac{1}{x}\right)^3 = x^3 + 3x + \frac{3}{x} + \frac{1}{x^3} = \left(x^3+\frac{1}{x^3}\right) + 3\left(x+\frac{1}{x}\right),$$

giving

$$x^3 + \frac{1}{x^3} = y^3 - 3y.$$

In this case,

$$(y^3 - 3y)^2 = \left(x^3+\frac{1}{x^3}\right)^2 = \left(x^6+\frac{1}{x^6}\right) + 2$$

and

$$\left(x^6+\frac{1}{x^6}\right) = (y^3 - 3y)^2 - 2 = y^6 - 6y^4 + 9y^2 - 2.$$

Thus the given function is

$$\frac{y^6 - (y^6 - 6y^4 + 9y^2 - 2) - 2}{y^3 + (y^3 - 3y)} = \frac{3y(2y^3 - 3y)}{2y^3 - 3y} = 3y,$$

provided $2y^3 \neq 3y$.

Clearly $y = x + \frac{1}{x}$ is never zero for $x > 0$, and so this condition reduces to

$$2y^2 \neq 3.$$

Now, for

$$2y^2 = 2\left(x^2 + 2 + \frac{1}{x^2}\right) = 3,$$

then

$$2x^2 + 1 + \frac{2}{x^2} = 0,$$

$$2x^4 + x^2 + 2 = 0,$$

with discriminant -15, implying a nonreal value of x. Hence, for $x > 0$, $2y^2$ is never equal to 3, and the value of the function is simply

$$3y = 3\left(x + \frac{1}{x}\right).$$

Since it is well known that the minimum value of $x + \frac{1}{x}$ is 2 (occurring when $x = 1$), the minimum of the given function is 6, and occurs for $x = 1$.

6. $\triangle DXY$ is inscribed in rectangle $ABCD$ as shown in the following figure. If the areas of triangles AXD, BXY, and DYC, respectively, are 5, 4, and 3 units, what is the area of $\triangle DXY$?

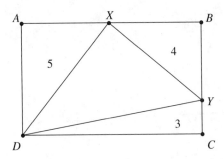

Let the lengths of the parts of the sides be represented as shown in Figure 3.

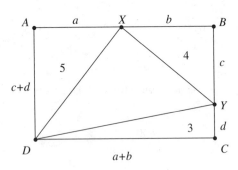

FIGURE 3

Then from the given areas we easily have

(i) $a(c + d) = 10$,
(ii) $bc = 8$, and
(iii) $d(a + b) = 6$.

Solutions To The Challenges

We aren't required to solve these equations for a, b, c, d. Our goal is to determine the value of $\triangle DXY = ABCD - (5 + 4 + 3) = ABCD - 12$, which is equivalent to finding the area R of $ABCD$.

From the above equations,

$$R = (a+b)(c+d)$$
$$= \frac{6}{d} \cdot \frac{10}{a} = \frac{60}{ad}.$$

The problem, then, reduces to finding ad, which it is not at all clear how to do.

Let the figure be decomposed by lines parallel to the sides of the rectangle as shown in Figure 4.

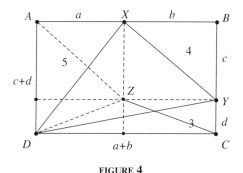

FIGURE 4

Then

$$\triangle DXY = \triangle XYZ + \triangle DXZ + \triangle DYZ$$
$$= 4 + \triangle AXZ + \triangle CYZ$$
$$= 4 + \tfrac{1}{2}ac + \tfrac{1}{2}bd,$$

and

$$R = \triangle DXY + 12 = 16 + \tfrac{1}{2}ac + \tfrac{1}{2}bd,$$

that is,

$$R = 16 + \tfrac{1}{2}(ac + bd).$$

Also

$$R = (a+b)(c+d) = ac + ad + bc + bd,$$

and so we have

$$ac + ad + bc + bd = 16 + \tfrac{1}{2}(ac + bd),$$

giving
$$\tfrac{1}{2}(ac + bd) + ad = 8 \quad (\text{recall } bc = 8).$$

Thus
$$\tfrac{1}{2}(ac + bd) = 8 - ad,$$

in which case,
$$R = 16 + \tfrac{1}{2}(ac + bd) = 24 - ad.$$

Combining with the earlier $R = 60/ad$, we get
$$\frac{60}{ad} = 24 - ad,$$
$$60 = 24ad - (ad)^2,$$
$$(ad)^2 - 24ad + 60 = 0,$$

and
$$ad = \frac{24 \pm \sqrt{24^2 - 240}}{2}$$
$$= 12 \pm 2\sqrt{21}.$$

Finally, then,
$$R = 24 - (12 \pm 2\sqrt{21}) = 12 \pm 2\sqrt{21}.$$

Since $R = 5 + 4 + 3 + \triangle DXY > 12$, we must take the $+$ sign here, giving
$$R = 12 + 2\sqrt{21}, \quad \text{and} \quad \triangle DXY = 2\sqrt{21}.$$

7. What is the area of the shaded region in the following figure?

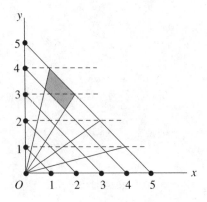

Referring to Figure 5, where the shaded region is labelled s, let the neighboring regions be labelled t, u, v, as shown.

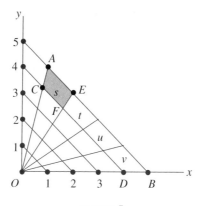

FIGURE 5

Clearly triangles AOB and COD are similar, and the ratio of corresponding sides is

$$\frac{OD}{OB} = \frac{4}{5}.$$

Therefore, since the areas of similar triangles are proportional to the squares on corresponding sides,

$$\triangle COD = \frac{16}{25} \triangle AOB.$$

It follows, then, that their difference

$$s + t + u + v = \frac{9}{25} \triangle AOB.$$

Similarly,

$$\triangle FOD = \frac{16}{25} \triangle EOB, \quad \text{and} \quad t + u + v = \frac{9}{25} \triangle EOB,$$

and we have

$$\begin{aligned} s &= \frac{9}{25}(\triangle AOB - \triangle EOB) \\ &= \frac{9}{25}\left(\frac{1}{2} \cdot 5 \cdot 4 - \frac{1}{2} \cdot 5 \cdot 3\right) \\ &= \frac{9}{25} \cdot \frac{1}{2} \cdot 5 = \frac{9}{10}. \end{aligned}$$

8. There are ten prizes, five A's, three B's, and two C's, placed in identical sealed envelopes for the top ten contestants in a mathematics contest. The prizes are awarded by allowing winners to select an envelope at random from those remaining. When the eighth contestant goes to select a prize, what is the probability that the remaining three prizes are one A, one B, and one C?

The last three envelopes are an unordered subset of three of the ten envelopes. Thus there are $\binom{10}{3} = 120$ sets of three envelopes that can occur as the last three.

The number of unordered subsets of three envelopes that contain one A, one B, and one C is $5 \cdot 3 \cdot 2 = 30$: any of the five A's may occur with any of the three B's and either of the two C's.

Therefore the required probability is

$$\frac{30}{120} = \frac{1}{4}.$$

9. (Two exercises from *Quantum*, Vol. 7, July/Aug, 1997)

   (a) What is the minimum value of the function

   $$f(x) = \frac{x^2}{8} + x \cos x + \cos 2x?$$

   (b) Prove that the equation $x^3 - x - 3 = 0$ has a unique real root and that its value is greater than $\sqrt[5]{13}$.

(a) Substituting $2\cos^2 x - 1$ for $\cos 2x$, we get

$$f(x) = \frac{x^2}{8} + x \cos x + 2 \cos^2 x - 1$$

$$= \frac{1}{8}(x^2 + 8x \cos x + 16 \cos^2 x) - 1$$

$$= \frac{1}{8}(x + 4 \cos x)^2 - 1,$$

revealing that the minimum is -1 when $\cos x = -x/4$.

(b) Let $f(x) = x^3 - x - 3$.

Since $f(1) = -3 < 0$ and $f(2) = 3 > 0$, the equation has a real root a between 1 and 2. Let the other roots be b and c. Then

$$f(x) = x^3 - x - 3$$
$$= (x-a)(x-b)(x-c)$$
$$= (x-a)[x^2 - (b+c)x + bc],$$

and in order to show a is a unique real root we need to show that the discriminant of the quadratic factor here is negative, i.e.,

$$(b+c)^2 < 4bc.$$

Now, from the equation we know that $a+b+c = 0$ and $abc = 3$. Hence

$$b + c = -a \quad \text{and} \quad bc = \frac{3}{a},$$

and the required $(b+c)^2 < 4bc$ requires $a^2 < 4 \cdot \frac{3}{a}$, or $a^3 < 12$, which is clearly so since $a < 2$. Thus a is a unique real root.

Since a is a root, it satisfies the equation, giving $a^3 - a - 3 = 0$. Thus $a^3 = a + 3$, and

$$a^5 = a^2 \cdot a^3 = a^2(a+3)$$
$$= a^3 + 3a^2 = (a+3) + 3a^2$$
$$= 3a^2 + a + 3.$$

We would like to show that $a^5 > 13$, i.e.,

$$3a^2 + a + 3 > 13,$$
$$3a^2 + a - 10 > 0,$$
$$(3a - 5)(a + 2) > 0,$$

which clearly holds for $a > \frac{5}{3}$. It remains, then, only to show that $a > \frac{5}{3}$.

Now, a lies between 1 and 2, and if $f(\frac{5}{3})$ is negative, it would follow that a lies between $\frac{5}{3}$ and 2, giving the desired $a > \frac{5}{3}$. Hence we need only check to see that $f(\frac{5}{3}) < 0$:

$$f\left(\frac{5}{3}\right) = \left(\frac{5}{3}\right)^3 - \frac{5}{3} - 3 = \frac{1}{27}(125 - 45 - 81) = -\frac{1}{27},$$

and the solution is complete.

10. (Problem B214 from *Quantum*, Vol. 8, Sept/Oct, 1997)

Find the fraction $\frac{a}{b}$ with smallest denominator that lies between $\frac{97}{36}$ and $\frac{96}{35}$:

$$\frac{97}{36} < \frac{a}{b} < \frac{96}{35}.$$

Multiplying through by b, we get

$$\frac{97}{36}b < a < \frac{96}{35}b,$$

that is,

$$2.694b < a < 2.743b, \quad \text{approximately}.$$

Obviously there is no integer between 2.694 and 2.743 themselves, and so b must exceed 1. Since $2.743 - 2.694 = 0.049$, which is slightly more than $\frac{1}{21}$, for $b = 21$ the difference between $2.694\,b$ and $2.743\,b$ will exceed 1, and whatever their values, they can't help straddling an integer. Thus we are encouraged by the thought that b doesn't exceed 21.

However this seems to be the only comfort we're going to get. Sometimes one has to give up the deductive trail and resort to proceeding by trial. Mercifully we only have to go to $b = 7$, at which point

$$\frac{97}{36} \cdot 7 < 19 < \frac{96}{35} \cdot 7,$$

giving the desired fraction to be $\frac{19}{7}$.

11. (Problem 637 from *The College Mathematics Journal*, 29, Nov., 1998)

The sequence $\{a_n\}$ is defined by

$$a_1 = \sqrt{2}, \quad \text{and for } n \geq 1, \quad a_{n+1} = \sqrt{2 + a_n}.$$

Determine

$$\lim_{n \to \infty} 4^n (2 - a_n).$$

(Solution by Ian McGee, University of Waterloo)

The problem unravels completely with the perceptive observation that

$$a_1 = \sqrt{2} = 2 \cdot \frac{1}{\sqrt{2}} = 2 \cos \frac{\pi}{4}.$$

Then
$$a_2 = \sqrt{2+a_1} = \sqrt{2+2\cos\frac{\pi}{4}} = \sqrt{2\left(1+2\cos^2\frac{\pi}{8}-1\right)} = 2\cos\frac{\pi}{8},$$
and in general, if
$$a_n = 2\cos\frac{\pi}{2^{n+1}},$$
then
$$a_{n+1} = \sqrt{2+2\cos\frac{\pi}{2^{n+1}}} = \sqrt{2\left(1+2\cos^2\frac{\pi}{2^{n+2}}-1\right)} = 2\cos\frac{\pi}{2^{n+2}}.$$

Hence, by induction,
$$a_n = 2\cos\frac{\pi}{2^{n+1}} \quad \text{for all } n.$$

Then
$$2 - a_n = 2 - 2\cos\frac{\pi}{2^{n+1}} = 2\left[1 - \left(1 - 2\sin^2\frac{\pi}{2^{n+2}}\right)\right] = 4\sin^2\frac{\pi}{2^{n+2}},$$
and
$$4^n(2-a_n) = 4^{n+1}\sin^2\frac{\pi}{2^{n+2}} = 4^{n+1}\left(\frac{\sin\frac{\pi}{2^{n+2}}}{\frac{\pi}{2^{n+2}}}\right)^2 \cdot \frac{\pi^2}{4^{n+2}}$$
$$= \frac{\pi^2}{4}\left(\frac{\sin\frac{\pi}{2^{n+2}}}{\frac{\pi}{2^{n+2}}}\right)^2.$$

Since the derivative of $\sin x$ exists at $x = 0$ and is equal to $\cos(0) = 1$, then $\lim_{n\to\infty} 4^n(2-a_n)$ exists, and its value is
$$\lim_{n\to\infty} 4^n(2-a_n) = \frac{\pi^2}{4} \cdot 1 = \frac{\pi^2}{4}.$$

12. (From the 1991 International Olympiad)

For any point P inside any $\triangle ABC$, show that at least one of the angles x, y, z does not exceed $30°$ (Figure 6).

This result follows easily from the simplest facts about Brocard geometry (for an introduction to Brocard geometry see my *Episodes in 19th and 20th Century Euclidean Geometry*, MAA, NML series, vol. 37, 1995).

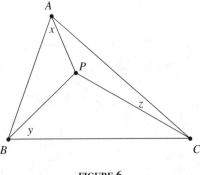

FIGURE 6

Recall that in each triangle there is a point Ω, a so-called Brocard point, such that the three angles ΩAB, ΩBC, ΩCA are equal (see Figure 7). Their common value is denoted by ω and is called the Brocard angle of the triangle.

Now, it is known that the Brocard angle ω is never greater than $30°$. For P at Ω, then, $x = y = z = \omega \leq 30°$; and since any other point P inside $\triangle ABC$ lies in one of the three sub-triangles $A\Omega B$, $B\Omega C$, $C\Omega A$, it is clear for such a point that at least one of x, y, z is less than ω (see Figure 7).

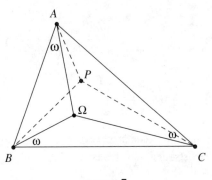

FIGURE 7

13. In Figure 8, $ABCD$ is a cyclic quadrilateral constructed on diameter AC as follows. B is chosen arbitrarily on one of the semicircular arcs AC. M is the midpoint of AB and the point P is taken on CM such that $MP = AM$. Extending BP then gives D on the circumference. Finally, CQ is drawn parallel to AP.

Prove $BP = QD$ for all positions of B.

Referring to Figure 9, let $\angle BAP = x$, $\angle PAC = y$, and $\angle ACB = z$. Since AC is a diameter, the large angles at B and D are right angles. Since P lies on

Solutions To The Challenges

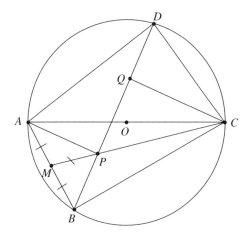

FIGURE 8

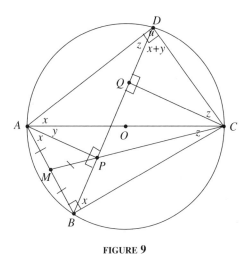

FIGURE 9

the semicircle with center M and radius one-half AB, $\angle APB$ is a right angle, and since CQ is parallel to AP, then QC is also perpendicular to BD.

In $\triangle ABP$, x is the complement of $\angle ABP$, as is $\angle PBC$ at the point B. Thus $\angle PBC = x$, and since angles in the same segment are equal, we also have $\angle CAD = x$.

In $\triangle ABC$ we have $x + y + z = 90°$. Since $\angle BDC = \angle BAC = x + y$ in the same segment, then in triangles APD and DQC the angles at D and C are z.

Now clearly

$$\tan z = \frac{AB}{BC} = \frac{AP}{PD}.$$

Also

$$\cos x = \frac{AP}{AB} = \frac{BQ}{BC}, \quad \text{implying} \quad \frac{AB}{BC} = \frac{AP}{BQ}.$$

Hence

$$\frac{AP}{PD} = \frac{AP}{BQ}, \quad \text{and} \quad PD = BQ.$$

Thus

$$BP = BQ - PQ = PD - PQ = QD.$$

14. (Seven re-phrased problems from the 1999 AIME [American Invitational Mathematics Examination])

(i) A transformation of the first quadrant of the coordinate plane maps each point (x, y) to the point $(x', y') = (\sqrt{x}, \sqrt{y})$. The vertices of the quadrilateral $ABCD$ are $A = (900, 300)$, $B = (1800, 600)$, $C = (600, 1800)$, and $D = (300, 900)$. What is the area of the region enclosed by the image of $ABCD$?

The images of A, B, C, D, and are easily calculated:

$A = (900, 300)$, $B = (1800, 600)$,
$C = (600, 1800)$, $D = (300, 900)$;
$A' = (30, 10\sqrt{3})$, $B' = (30\sqrt{2}, 10\sqrt{6})$,
$C' = (10\sqrt{6}, 30\sqrt{2})$, $D' = (10\sqrt{3}, 30)$.

From the transformation we have $x' = \sqrt{x}$, giving $(x')^2 = x$; similarly $(y')^2 = y$. Now, the equation of the line AB is $x = 3y$, and so the image of the segment AB lies on the locus whose equation is $(x')^2 = 3(y')^2$, which is just the pair of straight lines $x' = \pm\sqrt{3}y'$ through the origin. Since $\sqrt{x}$ and $\sqrt{y}$ are always nonnegative, the appropriate sign here is $+$. Thus the image of the segment AB is the segment on the line $x' = \sqrt{3}y'$ between the images A' and B' (whose coordinates clearly satisfy this equation). By the same reasoning, the image of any segment on a line through the origin is a segment on a line through the origin. Thus the image of CD is the segment $C'D'$.

However, it is different with the images of BC and AD. The equation of the line BC is evidently $x + y = 2400$, which yields $(x')^2 + (y')^2 = 2400$

in the $x'y'$ image plane. Thus the image of the segment BC is the arc of this circle which lies between B' and C'. Similarly, the image of the segment AD is the arc of the circle $(x')^2 + (y')^2 = 1200$ between A' and D', making the image of $ABCD$ a section of an annulus (Figure 10).

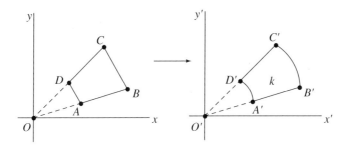

FIGURE 10

From the coordinates of A', we have

$$\tan \angle x'O'A' = \frac{30}{10\sqrt{3}} = \frac{1}{\sqrt{3}},$$

implying $\angle x'O'A' = 30°$. From D' we similarly get $\angle D'O'y' = 30°$, making $\angle A'O'D' = 30°$. Accordingly, the image of $ABCD$ is $\frac{1}{12}$ of the annulus between the circles, center the origin, and radii $\sqrt{2400}$ and $\sqrt{1200}$. Hence the required area is

$$\tfrac{1}{12}(2400\pi - 1200\pi) = 100\pi.$$

(ii) There is a set of 1000 switches, each of which has four positions, called A, B, C, D. When the position of any switch changes, it is only from A to B, from B to C, from C to D, or from D to A. Initially each switch is in position A. The switches are labeled with the 1000 different integers $2^x 3^y 5^z$, where each of x, y, z takes the values $0, 1, \ldots, 9$. Moreover, the switches are numbered switch 1, switch 2, ..., switch 1000.

At step i of a 1000-step process, the ith switch is advanced one step, and so are all the other switches whose labels divide the label on the ith switch. After step 1000 has been completed, how many switches will be in position A?

The switch labeled $2^a 3^b 5^c$ is turned for each triple (x, y, z) in which

$$x \geq a, \quad y \geq b, \quad \text{and} \quad z \geq c,$$

that is, for

x equal to any of the $9 - (a - 1) = 10 - a$ values $\{a, a+1, \ldots, 9\}$,
y equal to any of the $(10 - b)$ values $\{b, b+1, \ldots, 9\}$,
and
z equal to any of the $(10 - c)$ values $\{c, c+1, \ldots, 9\}$.

Thus the switch labeled $2^a 3^b 5^c$ is turned a total of $N = (10-a)(10-b) \cdot (10 - c)$ times, and ends in position A if and only if N is divisible by 4. Observing that

the number of switches for which 4 divides N
$= 1000 - $ (the number of switches for which 4 does **not** divide N),

the desired number is easily calculated as follows.

If two or more of the factors in N are even, N will be divisible by 4. Hence there are only two cases when N is not divisible by 4:

(i) each factor in N is odd,
(ii) two of the factors in N are odd and the third is divisible by 2 but not by 4; that is, when N is just singly even.

Since there are five odd and five even numbers among the values 0, 1, ..., 9 which are taken by a, b, c, there are five ways to make a factor of N odd, implying N is odd $5^3 = 125$ times.

Now, a singly even factor of N can only be 2, 6, or 10 (obviously 4 and 8 must be avoided). Thus there are $3 \cdot 5 \cdot 5 = 75$ ways of making a specified factor singly even and the other two odd. Since any of the three factors could be the even one, N is singly even $3 \cdot 75 = 225$ ways, for a total of $125 + 225 = 350$ times that N is not divisible by 4. Hence the number of switches that end in position A is

$$1000 - 350 = 650.$$

(iii) Let T be the set of ordered triples (x, y, z) of nonnegative real numbers that lie in the plane $x + y + z = 1$. Let us say that (x, y, z) **supports** (a, b, c) when exactly two of the following are true:

$$x \geq a, \quad y \geq b, \quad z \geq c.$$

Let S consist of those triples in T that support the point $P(\frac{1}{2}, \frac{1}{3}, \frac{1}{6})$. Determine the ratio

$$\frac{\text{area of } S}{\text{area of } T}.$$

Clearly each of the points $A(1, 0, 0)$, $B(0, 1, 0)$, $C(0, 0, 1)$ lies on the plane $x + y + z = 1$, implying that T is just the equilateral triangle ABC in the nonnegative octant which has side $s = \sqrt{2}$ and area

$$\Delta = \frac{s^2 \sqrt{3}}{4} = \frac{\sqrt{3}}{2} \quad \text{(Figure 11)}.$$

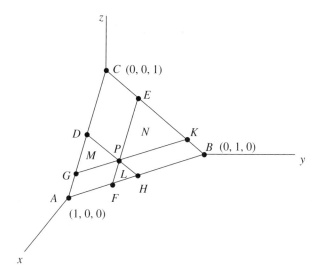

FIGURE 11

Since $\frac{1}{2} + \frac{1}{3} + \frac{1}{6} = 1$, we observe that the point P lies in T. Now, the planes $x = \frac{1}{2}, y = \frac{1}{3}, z = \frac{1}{6}$ all go through P and divide T into three parallelograms and three triangles L, M, N, where

L is the triangle in which $\left(x \geq \frac{1}{2}, y \geq \frac{1}{3}, z < \frac{1}{6}\right)$,

M is the triangle where $\left\{x \geq \frac{1}{2}, y < \frac{1}{3}, z \geq \frac{1}{6}\right\}$,

and

N is the triangle such that $\left\{x < \frac{1}{2}, y \geq \frac{1}{3}, z \geq \frac{1}{6}\right\}$.

Thus we have $S = L \cup M \cup N$.

The sum of the areas of the triangles L, M, and N is found by subtracting from T the areas of the three parallelograms:

$$\text{Area of } S = T - (PGAF + PHBK + PECD).$$

Now,

$$\text{parallelogram } PGAF = 2 \cdot \triangle AFG,$$

and since $AF = \frac{1}{3}AB = \frac{\sqrt{2}}{3}$, $AG = \frac{1}{6}AC = \frac{\sqrt{2}}{6}$, and $\angle A = 60°$, we have

$$\text{parallelogram } PGAF = 2\left(\frac{1}{2} \cdot \frac{\sqrt{2}}{3} \cdot \frac{\sqrt{2}}{6} \cdot \sin 60°\right)$$

$$= 2\left(\frac{1}{2} \cdot \frac{\sqrt{2}}{3} \cdot \frac{\sqrt{2}}{6} \cdot \frac{\sqrt{3}}{2}\right) = \frac{\sqrt{3}}{18}.$$

Similarly,

$$PHBK = 2\left(\frac{1}{2} \cdot \frac{\sqrt{2}}{2} \cdot \frac{\sqrt{2}}{6} \cdot \frac{\sqrt{3}}{2}\right) = \frac{\sqrt{3}}{12},$$

and

$$PECD = 2\left(\frac{1}{2} \cdot \frac{\sqrt{2}}{2} \cdot \frac{\sqrt{2}}{3} \cdot \frac{\sqrt{3}}{2}\right) = \frac{\sqrt{3}}{6}.$$

Hence

$$\text{the area of } S = \frac{\sqrt{3}}{2} - \sqrt{3}\left(\frac{1}{18} + \frac{1}{12} + \frac{1}{6}\right)$$

$$= \frac{\sqrt{3}}{36}(18 - 2 - 3 - 6) = \frac{7\sqrt{3}}{36},$$

and the required ratio is

$$\frac{\frac{7\sqrt{3}}{36}}{\frac{\sqrt{3}}{2}} = \frac{7}{18}.$$

(iv) Ten points in the plane are given, with no three collinear. Four distinct segments joining pairs of these points are chosen at random, all such segments being equally likely. What is the probability that some three of the four chosen segments form a triangle whose vertices are among the ten given points?

Solutions To The Challenges

The number of segments is
$$\binom{10}{2} = 45,$$
the number of triangles is
$$\binom{10}{3} = \frac{10 \cdot 9 \cdot 8}{2 \cdot 3} = 120,$$
and the number of ways of selecting four segments is $\binom{45}{4}$.

Now, each triangle can be combined with each of the other 42 segments to give a quadruple of four segments, for a total of $120 \cdot 42$ favourable selections. Thus the required probability is
$$\frac{120 \cdot 42}{\frac{45 \cdot 44 \cdot 43 \cdot 42}{2 \cdot 3 \cdot 4}} = \frac{16}{473}.$$

(v) Find the acute angle θ, given that $\sum_{k=1}^{35} \sin 5k = \tan \theta$, where the angles are measured in degrees.

Let
$$S = \sum_{k=1}^{35} \sin 5k = \sin 5 + \sin 10 + \cdots + \sin 175.$$
Then
$$S \cos 5 = \sin 5 \cdot \cos 5 + \sin 10 \cdot \cos 5 + \sin 15 \cdot \cos 5 + \cdots$$
$$+ \sin 170 \cdot \cos 5 + \sin 175 \cdot \cos 5.$$
Since $\sin A \cos B = \frac{1}{2}[\sin(A+B) + \sin(A-B)]$, it follows that
$$S \cos 5 = \frac{1}{2}\big[(\sin 10 + \sin 0) + (\sin 15 + \sin 5)$$
$$+ (\sin 20 + \sin 10) + \cdots + (\sin 180 + \sin 170)\big],$$
giving
$$2S \cos 5 = \sin 5 + \sin 10 + \sin 15 + \cdots + \sin 170$$
$$+ \sin 10 + \sin 15 + \cdots + \sin 170 + \sin 175$$
$$= 2(\sin 5 + \sin 10 + \sin 15 + \cdots + \sin 170)$$
since $\sin 5 = \sin 175$. Hence

$$S\cos 5 = \sin 5 + \sin 10 + \sin 15 + \cdots + \sin 170$$
$$= S - \sin 175$$
$$= S - \sin 5.$$

Thus
$$S(1 - \cos 5) = \sin 5,$$

giving
$$S = \frac{\sin 5}{1 - \cos 5} = \frac{(1 + \cos 5) \cdot \sin 5}{1 - \cos^2 5}$$
$$= \frac{(1 + \cos 5) \cdot \sin 5}{\sin^2 5} = \frac{1 + \cos 5}{\sin 5}.$$

In Figure 12, then, we have
$$S = \frac{1 + \frac{b}{c}}{\frac{a}{c}} = \frac{c + b}{a} = \tan \angle DBC.$$

In right triangle ABC, $\angle ABC = 85$, and since $\triangle DAB$ is isosceles, we have $2x = 5$, making $\angle DBC = \theta = 87\frac{1}{2}$ degrees.

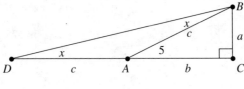

FIGURE 12

(vi) Forty teams play a tournament in which every team plays every other team exactly once. No ties occur, and each team has a 50% chance of winning any game it plays. Determine the probability that no two teams win the same number of games.

(Solution by Ron Dunkley, University of Waterloo)

Let the teams be numbered $T_1, T_2, \ldots, T_{40}$. The result of T_i's play in the tournament, then, may be recorded in an ordered 40-tuple $R_i = (1, 0, \ldots, \_\_, \ldots)$, having a blank in component i since T_i does not play itself, but otherwise whose kth component is 0 if T_i lost to T_k and 1 if T_i beat T_k. The outcome of the entire tournament is similarly given by an ordered 40-tuple

Solutions To The Challenges **231**

$$W = \{w_1, w_2, \ldots, w_{40}\}$$

where w_i is the number of wins by team T_i.

Since each team plays the other 39 teams, it follows that w_i is confined to the values $0, 1, \ldots, 39$, and in the event that no two teams win the same number of games, the components of W must be these 40 different integers in some order. Clearly, then, there are 40! ways in which this outcome can be realized, and the probability that no two teams win the same number of games is 40! times the probability that any one of these "all different" outcomes occurs. Accordingly, then, let us determine the probability of the particular outcome

$$W = \{39, 38, 37, \ldots, 1, 0\}.$$

In this case, the 40 by 40 "matrix" obtained by stacking the records R_i is

$$R_1 = \{\_, 1, 1, 1, \ldots, 1, 1\},$$
$$R_2 = \{0, \_, 1, 1, \ldots, 1, 1\},$$
$$R_3 = \{0, 0, \_, 1, \ldots, 1, 1\},$$
$$\vdots$$
$$R_{40} = \{0, 0, 0, 0, \ldots, 0, \_\}.$$

Since T_1 won all its games, R_1 contains 39 1's, and the probability of this happening is $(\frac{1}{2})^{39}$. Now, a win for one team is a loss for another. Thus T_1's 39 wins not only fills R_1 with 1's but fills the first column of the matrix with 0's.

Coming to R_2, then, there are only 38 undeclared components. The probability of all these being 1's, with their 38 corresponding 0's that complete the second column, is $(\frac{1}{2})^{38}$.

Similarly, this leaves 37 unspecified components in R_3, and the probability they are all 1's, with their corresponding 0's that complete the third column, is $(\frac{1}{2})^{37}$. Continuing in this way we obtain the probability of attaining the above matrix of results to be

$$\left(\frac{1}{2}\right)^{39} \cdot \left(\frac{1}{2}\right)^{38} \cdot \left(\frac{1}{2}\right)^{37} \cdots \left(\frac{1}{2}\right)^{0} = \frac{1}{2^{780}}.$$

Hence the probability that no two teams won the same number of games is

$$\frac{40!}{2^{780}}.$$

(vii) Consider the paper triangle whose vertices are $(0, 0)$, $(34, 0)$, and $(16, 24)$. The vertices of its medial triangle are the midpoints of its sides. A triangular pyramid is formed by folding the triangle along the sides of its medial triangle. What is the volume of this pyramid?

It is well known that the medial triangle divides the given triangle into four congruent sub-triangles and that the sides of the medial triangle are one-half the sides of the given triangle. Thus, in Figure 13, the lengths of the edges of the base of the pyramid are

$$EF = \tfrac{1}{2}OA = 17,$$

$$DE = \tfrac{1}{2}OB = \tfrac{1}{2}\sqrt{16^2 + 24^2} = \tfrac{1}{2}\sqrt{832} = \sqrt{208},$$

and

$$DF = \tfrac{1}{2}AB = \tfrac{1}{2}\sqrt{18^2 + 24^2} = 15.$$

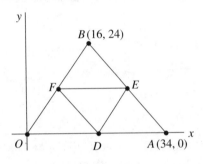

FIGURE 13

Also, the area of the base of the pyramid is

$$\triangle DEF = \tfrac{1}{4}\triangle ABC = \tfrac{1}{4}\left(\tfrac{1}{2} \cdot 34 \cdot 24\right) = 102.$$

Now, as in Figure 14, let the pyramid be placed in a three-dimensional Cartesian frame of reference with E at the origin, its base $\triangle DEF$ in the xy plane with F at $(17, 0, 0)$, D at $(a, b, 0)$, and its fourth vertex G at (x, y, z).

Since the volume V of a pyramid is $\tfrac{1}{3} \cdot$ area of the base $\cdot$ the altitude, we have

$$V = \tfrac{1}{3} \cdot 102 \cdot z = 34z,$$

and it remains to determine the altitude z. From the known lengths of the edges of the pyramid, this is a straightforward calculation.

(1) $GE = 15$, giving $x^2 + y^2 + z^2 = 225$,
(2) $GF = \sqrt{208}$, giving $(x - 17)^2 + y^2 + z^2 = 208$, from which

Solutions To The Challenges

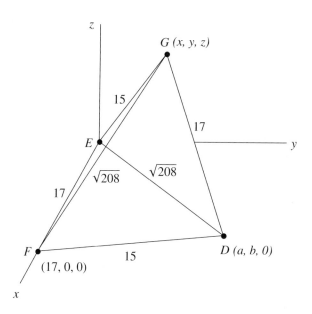

FIGURE 14

$$34x - 289 = 17, \quad 2x - 17 = 1, \quad \text{and} \quad x = 9$$

and

$$y^2 + z^2 = 225 - 81 = 144.$$

(3) $ED = \sqrt{208}$, giving $a^2 + b^2 = 208$,
(4) $FD = 15$, giving $(a - 17)^2 + b^2 = 225$, from which

$$34a - 289 = -17, \quad \text{and} \quad a = 8, \quad \text{and} \quad b = 12.$$

(5) $DG = 17$, giving

$$(x - a)^2 + (y - b)^2 + z^2 = 289, \quad 1 + (y - 12)^2 + z^2 = 289.$$

Since $y^2 + z^2 = 144$, then, $-24y + 144 = 144$, and $y = 0$. Surprisingly, G lies in the xz coordinate plane.

Finally,

$$z^2 = 144, \quad z = 12,$$

and

$$V = 34z = 34 \cdot 12 = 408.$$

15. Prove the remarkable discoveries of Jakob Bernoulli (1654–1705) that

(a) $\sum_{k=1}^{\infty} \dfrac{k^2}{2^k} = 6$ and (b) $\sum_{k=1}^{\infty} \dfrac{k^3}{2^k} = 26.$

In Solution 2 we approach the problem directly by taking the limit of partial sums. This is a straightforward procedure that gives the result without difficulty. However, let us begin with a very slick approach that was proposed by Robert Burckel of Kansas State University.

Solution 1:

(a) Consider the geometric series

$$\sum_{k=1}^{\infty} \frac{x^k}{2^k} = \frac{\frac{x}{2}}{1 - \frac{x}{2}} = \frac{x}{2-x}.$$

This converges for $0 \leq \frac{x}{2} < 1$, in particular at $x = 1$, and therefore the series obtained by differentiating it term-by-term also converges at $x = 1$. Differentiating gives

$$\sum_{k=1}^{\infty} \frac{kx^{k-1}}{2^k} = \frac{2}{(2-x)^2},$$

and multiplying by x, we get

$$\sum_{k=1}^{\infty} \frac{kx^k}{2^k} = \frac{2x}{(2-x)^2}.$$

A second differentiation yields another series that converges at $x = 1$, and we have

$$\sum_{k=1}^{\infty} \frac{k^2 x^{k-1}}{2^k} = \frac{2x + 4}{(2-x)^3},$$

which gives the desired result upon setting $x = 1$.

The result in part (b) is obtained similarly by a second multiplication by x and another differentiation.

Solution 2:

(a) Let

$$S_n = \frac{1^2}{2^1} + \frac{2^2}{2^2} + \frac{3^2}{2^3} + \cdots + \frac{n^2}{2^n}.$$

Then
$$\frac{1}{2}S_n = \frac{1^2}{2^2} + \frac{2^2}{2^3} + \cdots + \frac{(n-1)^2}{2^n} + \frac{n^2}{2^{n+1}},$$
and subtraction gives
$$\frac{1}{2}S_n = \frac{1^2}{2^1} + \frac{2^2 - 1^2}{2^2} + \frac{3^2 - 2^2}{2^3} + \cdots + \frac{n^2 - (n-1)^2}{2^n} - \frac{n^2}{2^{n+1}},$$
i.e.,
$$\frac{1}{2}S_n = \frac{1}{2} + \frac{3}{2^2} + \frac{5}{2^3} + \cdots + \frac{2n-1}{2^n} - \frac{n^2}{2^{n+1}}.$$

Similarly, multiplying again by $\frac{1}{2}$, we get
$$\frac{1}{4}S_n = \frac{1}{2^2} + \frac{3}{2^3} + \cdots + \frac{2n-3}{2^n} + \frac{2n-1}{2^{n+1}} - \frac{n^2}{2^{n+2}},$$
and subtraction gives
$$\frac{1}{4}S_n = \frac{1}{2} + \frac{2}{2^2} + \frac{2}{2^3} + \cdots + \frac{2}{2^n} - \frac{n^2 + 2n - 1}{2^{n+1}} + \frac{n^2}{2^{n+2}}$$
$$= \frac{1}{2} + \left[\frac{1}{2} + \frac{1}{2^2} + \cdots + \frac{1}{2^{n-1}}\right] - \frac{n^2 + 2n - 1}{2^{n+1}} + \frac{n^2}{2^{n+2}}$$
$$= \frac{1}{2} + \frac{\frac{1}{2}\left[1 - \left(\frac{1}{2}\right)^{n-1}\right]}{1 - \frac{1}{2}} - \frac{n^2 + 2n - 1}{2^{n+1}} + \frac{n^2}{2^{n+2}}$$
$$= \frac{1}{2} + \left[1 - \frac{1}{2^{n-1}}\right] - \frac{n^2 + 2n - 1}{2^{n+1}} + \frac{n^2}{2^{n+2}}$$
$$= \frac{3}{2} - \frac{1}{2^{n-1}} - \frac{n^2 + 2n - 1}{2^{n+1}} + \frac{n^2}{2^{n+2}},$$
giving
$$S_n = 6 - \frac{1}{2^{n-3}} - \frac{n^2 + 2n - 1}{2^{n-1}} + \frac{n^2}{2^n}.$$
Hence
$$\sum_{k=1}^{\infty} \frac{k^2}{2^k} = \lim_{n \to \infty} S_n = 6,$$
since, in the limit, the last three terms vanish, each being a polynomial divided by an exponential function.

(b) $\sum_{k=1}^{\infty} \frac{k^3}{2^k}$ is summed in the same way, except that one has to (multiply by $\frac{1}{2}$ and subtract) an extra time, with the result that

$$\frac{S_n}{8} = \frac{13}{4} + \text{terms which vanish in the limit,}$$

from which the desired result follows directly.

Index of Publications

The section(s) in which the publication is mentioned is given immediately after its name.

1. *Which Way Did The Bicycle Go?* by Joe Konhauser, Dan Velleman, and Stan Wagon, MAA Dolciani Series, Vol. 18, 1996: Sections 1, 11.
2. *Mathematical Miniatures* by Svetoslov Savchev and Titu Andreescu, The Anneli Lax New Mathematical Library Series, MAA: Sections 2, 17.
3. *The Two-Year College Mathematics Journal*, MAA: Sections 3, 4, 6, 9, 10, 12, 13, 16, 20.
4. *Problems and Solutions from the Mathematical Visitor*, 1877–1896, edited by Stanley Rabinowitz:, MathPro Press, 1991: Section 5.
5. *Introduction to Geometry*, H. S. M Coxeter, Wiley, 1961: Section 5.
6. *The Enjoyment of Mathematics*, Rademacher and Toeplitz, Princeton University Press, 1957: Section 5.
7. *The Problem Contest Book V* (the AHSME and AIME contests from 1983–1988), edited by George Berzsenyi and Stephen Maurer: Sections 5, 11.
8. *The Leningrad Mathematical Olympiads 1987–1991*, compiled by Dimitry Fomin and Alexey Kirichenko, MathPro Press, 1994: Section 5.
9. *Problem-Solving Through Problems* by Loren Larson, Springer-Verlag, 1983: Section 5.
10. *Graph Theory with Applications*, Bondy and Murty, American Elsevier, 1976: Section 6.
11. The magazine *Quantum*, National Science Teachers Association: Section 8.
12. *The American Mathematical Monthly*, MAA: Sections 8, 12.
13. *The $\pi\mu\varepsilon$ Journal*, The National Honorary Mathematics Society: Sections 11, 20.

14. *Ingenuity In Mathematics*, Ross Honsberger, The Anneli Lax New Mathematical Library Series, Volume 23, 1970: Section 11.
15. *Probability*, W. Feller, McGraw Hill, 1960: Section 12.
16. *Combinatorial Geometry in the Plane*, by Hadwiger, Debrunner, and Klee, Holt, Rinehart, and Winston, 1964: Section 13.
17. *Episodes from Nineteenth and Twentieth Century Euclidean Geometry*, Ross Honsberger, The Anneli Lax New Mathematical Library Series, Vol. 37, 1995: Section 15.
18. *Lure of the Integers*, Joe Rogers, Spectrum Series, MAA, 1992: Section 17.
19. *Matters Mathematical*, Herstein and Kaplansky, Harper and Row, 1974: Section 20.
20. *The Ontario Secondary School Mathematics Bulletin* The University of Waterloo: Section 21.
21. *Japanese Temple Geometry Problems* (in English), by Fukagawa and Pedoe, The Charles Babbage Research Centre, Winnipeg, Canada, 1989: Section 23.
22. The Journal *Gurukula Kangri Jijñana Patrika Aryabhata*: Section 23.
23. *The College Mathematics Journal*, MAA: Section 24.
24. *The Tournament of the Towns, 1980–1984*, edited by Peter Taylor, The Australian Mathematics Trust, 1993: Section 25.

Subject Index

It is hoped that these brief descriptions will identify a topic sufficiently to redirect you to its section. The contents are listed under the headings "Geometry and Combinatorial Geometry" and "Algebra, Number theory, and Probability." Subjects that are explicitly given in the title of a section are not listed.

Geometry and Combinatorial Geometry

	Section
Erdös' result on infinite sets and integral distances	4
Establish $\Delta = Rs$, where s = semiperimeter of the orthic triangle	5
Quartering a quadrilateral; segments to midpoints of sides	5
Dissecting a 2 by 10 rectangle to cover a square	5
Ellipse; given foci, tangent to x-axis; find major axis	5
A property of isosceles triangles	5
Reflecting a side of a triangle in the other two sides	5
Erdös' theorem on big distances in sets of diameter ≤ 1	6
Packing two squares into a third square	7
A locus of midpoints of segments between two circles	7
A certain set of equal circles in a square	8
The Schwab–Schoenberg mean	8
A property involving the median of a triangle	8
To construct an isosceles triangle to touch two given circles	8
On the triangle formed from the medians of a triangle	11
A construction problem on perpendicular medians	11
A variation on a problem of Regiomontanus	11
On the largest regular hexagon inscribed in a square	11
On circles capturing $\geq [n/3]$ points of a set of n planar points	11
Find lattice points on graph of $x^2 - 3xy - 13x + 5y + 11 = 0$	14
A certain rectangle and the relation $d^{2/3} - n^{2/3} = 1$	14
A special concurrency and collinearity	15
A property of cevian conjugates	15
A cevian property	15
The butterfly problem	16
The remarkable point Q	16
A property of the medial triangle	18
A property of regular n-gons inscribed in a unit circle	19
On a certain triangle of maximum area	21
A certain angle bisector in a parallelogram	21
On a certain square in a circle	21
On certain triangles which have a common Euler line	22
Subdividing a convex quadrilateral into n^2 sub-quadrilaterals	25

Algebra, Number Theory, and Probability

	Section
The footrace	5
A quadratic equation involving $[x]$, the integer part of x	5
An inequality on a sum of reciprocals and the rth prime	5
On prime decompositions and a perfect square	5
Polynomial; coefficients $a_0, a_n, f(1)$ odd implies no rational root	5
A certain quartic equation with an integer root	7
On multiples of 24	7
Quadratic $f(x)$ consists only of 1's when x consists only of 1's	8
The Schwab–Schoenberg mean	8
On $n = a_0 + a_1 \cdot 2 + a_2 \cdot 2^2 + \cdots + a_k \cdot 2^k, a_i \in \{0, 1, 2\}$	8
A property of sequences of positive integers	8
The fractions on either side of 5/8	8
The weight of watermelons	8
On certain fractions that reduce to integers	8
A set of three quations involving $x, [x], \{x\}$, etc.,	8
The strolling professor and her assistant	8
A sequence of positive integers $\{a_k\}$ with $k \mid a_1 + a_2 + \cdots + a_k$	8
Special properties of 30 and 24	9
On the functions $x^2 + y + 2$ and $y^2 + 4x$	11
On the minimum value of a certain function	11
On three cards marked with integers x, y, z	11
Mr. A thinks of a number	11
A property of 16-digit integers	11
Coin-tossing: odd man wins	12
On 16 positive integers with sum 100 and sum of squares 1000	14
Covering $1, 1/2, 1/3, \ldots$ with five bars	14
The greatest integer value of $\sqrt{x - 174} + \sqrt{x + 34}$	14
The value of $\sin\theta - \cos\theta$ for a certain θ	14
A property of uncountable sets of real numbers	18
The 37th digit in a certain decimal	20
An "oasis" property of prime numbers	20
Leapfrog induction	20
The method of infinite descent	20
Finding terms that are divisible by 11 in a certain sequence	22
A function f such that $f(i) \neq f(j)$ whenever $|i - j|$ is a prime	22

General Index

	Section
Algebraic integer	13
Arithmetic mean-geometric mean inequality	8
Analytic geometry	24
Binomial theorem	4
Brown, Steve	3
Bull, John	19
Butterfly problem	16
Cevian, cevian conjugate, cevian property, Ceva's theorem	15
Chebyshev polynomial	13
Circumcenter of a triangle	5, 22
Coloring lattice points	13
Complete graph	6
Countability of the rationals	18
De Moivre's theorem	4
Differentiation	7
Dilatation (Homothecy)	7, 22
Dinh, Hung	1
Dirichlet's theorem	20
Distance-rate-time problem	8
Euclid	19
Euler line	22
Farey sequences	8
Fejér, Leopold	5
Fractional part of an integer	8
Generating function	17
Graph, linear	3

	Section
Hahn, Liong-shin	11
Haruki, Hiroshi	16
Homothecy (Dilatation)	7, 22
Hyperbola	4
Induction	8, 10, 17, 20
Inequality concerning prime numbers	9
Infinite decimals	20
Infinite product	17
Inscribed circles	23
Integer distances	4
Interlocking sequences	8
Lambek, Joseph	17
Law of cosines	6
Law of sines	21
Linear dependence	5
Markov chain	12
Márquez, Juan-Bosco Romero	15
Matrix, $0-1$	5
Mendelsohn, Nathan	20
Menelaus' theorem	15
Minimum circle enclosing a set of points in the plane	11
Moser, Leo	17
Number of divisors of an integer	11
Orthic triangle	5, 22
Orthocenter of a triangle	22
Payan, Charles	1
Pythagorean triple	4
Rabinowitz, Stanley	5
Random points	3
Rank of a matrix	5
Recursion	8, 12
Reflection of the plane	5, 16
Reflector property of the ellipse	5
Relatively prime integers	9
Root-mean-square inequality	14

	Section
Rotation of coordinate axes	24
Rotation of the plane	5, 8
Sedinger, Harry	11
Sharygin, I. F.	8
Sierpinski, Waclaw	20
Silverman, David	4
Square-free part of an integer	5
Tanaka, Shotaro	23
Takeda, Shinko	23
Takeda, Shingen	23
Vector, geometric	13
Winterle, Riko	18

9693